AF595466

Liquid Crystal Optics for Applications

Liquid Crystal Optics for Applications

Editors

Kohki Takatoh

Akihiko Mochizuki

MDPI • Basel • Beijing • Wuhan • Barcelona • Belgrade • Manchester • Tokyo • Cluj • Tianjin

Editors
Kohki Takatoh
Department of Electrical Engineering, Faculty of Engineering, Sanyo-Onoda City University,
Sanyo-Onoda, Yamaguchi, Japan

Akihiko Mochizuki
Principal at i-CORE Technology, LLC.,
Louisville, CO, USA

Editorial Office
MDPI
St. Alban-Anlage 66
4052 Basel, Switzerland

This is a reprint of articles from the Special Issue published online in the open access journal *Crystals* (ISSN 2073-4352) (available at: https://www.mdpi.com/journal/crystals/special_issues/LC_Optics).

For citation purposes, cite each article independently as indicated on the article page online and as indicated below:

LastName, A.A.; LastName, B.B.; LastName, C.C. Article Title. *Journal Name* **Year**, *Volume Number*, Page Range.

ISBN 978-3-0365-6378-7 (Hbk)
ISBN 978-3-0365-6379-4 (PDF)

Contents

About the Editors

Kohki Takatoh

Kohki Takatoh has been a professor in the Department of Electrical Engineering, Sanyo-Onoda City University, former Tokyo University of Science, Yamaguchi, since 2005. He graduated from the Faculty of Science, Osaka University, and was awarded his MSC degree in 1983. In the same year, he joined the Toshiba Corporation. He was awarded a PhD degree for his thesis on liquid crystal research from Osaka University in 1992. From 1992 to 1994, he was a visiting Scholar at Hull University, supervised by Professor Goodby. In Toshiba Corporation, he was a leader of the liquid crystal development group. In 2000, he proposed "Viewing Angle Control (VAC) Filter" and launched a VAC Filter business in 2004. In 2005, VAC Filter was adopted by Toshiba Dynabook Tecra M3 and other applications. In Sanyo-Onoda City University, he proposed several types of new LCDs, including "Reverse TN-LCD (RTN)". He also studied new types of optical films using nanoparticles or LC monomers. He is a co-editor and co-author of "Handbook of Liquid Crystal". (2000) (Maruzen Co.) and co-author of "Alignment Technologies and Application of Liquid Crystal Devices" (2005) (Taylor & Francis Ltd.).

Akihiko Mochizuki

Dr. Akihiro Mochizuki was born in Japan. After working for Fujitsu Laboratories in Japan, since 1998, he has been in the US. He has over 42 years of research and development expertise, in both academia and industries. His research experience extends to both Japan and the US, and small entities and large entities. Based on his academic background, which is in low-dimensional electron conductive organic materials science, he has mainly been engaged in liquid crystal technologies, organic photo-conduction materials and their device applications. In these 25 years, he has been engaged in technical base management works based in Boulder, Colorado, US. His activity includes several scientific societies committee members, such as The International Society for Optics and Photonics (SPIE), Society for Information Display (SID) and so on. He has over 75 US patents and over 65 scientific publications, and 5 monograph publications, mainly in the liquid crystal research field.

Preface to "Liquid Crystal Optics for Applications"

After their discovery by Reinitzer in 1888, liquid crystals (LCs) were studied from the perspective of basic science for around 70 years. During this period, the researchers' interests were the fundamental relationship between the molecular structure and the liquid crystal properties, the physical properties such as elasticity and viscosity, the unique optical anisotropy, the response to electric or magnetic fields, the technologies for liquid crystal alignment on surfaces, etc. Due to these studies, in 1968, Heilmeir and his co-workers of RCA published the application of liquid crystal to a display device. After this invention, the industry of liquid crystal displays (LCDs) started in Japan. Since thin film transistors were applied to LCDs, the production of large LCDs could be realized. The LCD industry has been largely developed in Asia. Liquid crystal technology displays continue developing. However, the main interest regarding LC applications is moving to areas beyond displays.

LC devices control the light using an electric signal. A wide variety of applications are expected. Moreover, researchers can take advantage of display industry technologies. This Special Issue, "Liquid Crystal Optics for Applications", is timely. In this issue, 10 articles for new applications are collected. There is no overlap between the areas of the LC applications. This issue aims to provide an overview of the expansion of the research area for LC applications.

Technologies using liquid crystal on silicon (LCOS) have been investigated for many purposes. In this issue, three articles focus on LCOS usage for optical communication, laser processing and wireless optical charging. For other optical devices, we can find articles about wavelength selective filters and waveguides. We can find articles regarding technologies other than electronic devices that can be used for windows, and new functions for "smart windows" are shown. LC devices are effective not only for visible light but also for near-infrared radiation and terahertz waves. One article in this issue shows the LCD usage for terahertz waves. Research in the field of LC polymers has also been active. In one article of this issue, the application of LC polymers for polarizing grating is discussed. It is expected to be used in augmented reality (AR) technology, etc. LC material is also expected for applications without electrical effects. In this issue, we can find an article concerning LC's application to an ink, showing further possibilities regarding LC materials.

We want to express our sincere appreciation for Mr. Gilbert Liu, Section Managing Editor, MDPI, for his many important pieces of advice and contributions.

Kohki Takatoh and Akihiko Mochizuki
Editors

Article

Fast-Response Liquid Crystals for 6G Optical Communications

Junyu Zou [1], Zhiyong Yang [1], Chongchang Mao [2] and Shin-Tson Wu [1,*]

[1] College of Optics and Photonics, University of Central Florida, Orlando, FL 32816, USA; zoujunyuwinnie@Knights.ucf.edu (J.Z.); zhiyyang@knights.ucf.edu (Z.Y.)
[2] ElectroSience Lab, Ohio State University, 1330 Kinnear Road, Columbus, OH 43212, USA; mao.550@osu.edu
* Correspondence: swu@creol.ucf.edu; Tel.: +1-407-823-4763

Abstract: We report two high birefringence and low viscosity nematic mixtures for phase-only liquid-crystal-on-silicon spatial light modulators. The measured response time (on + off) of a test cell with 2π phase change at 1550 nm, 5 V operation voltage, and 40 °C is faster than 10 ms. To improve the photostability, a distributed Bragg reflector is designed to cutoff the harmful ultraviolet and blue wavelengths. These materials are promising candidates for future 6G optical communications.

Keywords: liquid crystals; spatial light modulator; liquid-crystal-on-silicon; photostability

Citation: Zou, J.; Yang, Z.; Mao, C.; Wu, S.-T. Fast-Response Liquid Crystals for 6G Optical Communications. *Crystals* **2021**, *11*, 797. https://doi.org/10.3390/cryst11070797

Academic Editors: Kohki Takatoh, Jun Xu and Akihiko Mochizuki

Received: 15 June 2021
Accepted: 6 July 2021
Published: 8 July 2021

1. Introduction

Dense wavelength division multiplexing (DWDM) optical networks [1] are the foundation of global internet communications. With an ability to transport over 100 information-bearing wavelengths on a single optical fiber [2], DWDM nullifies the cost-per-bit of communicating services. The technology makes it economically viable for businesses to move their computing and storage to the Cloud and for consumers to stream video ubiquitously. It is also an essential enabler for the 5th generation (5G) [3] and 6th generation [4,5] (6G) networks and most other advanced telecommunications. Ensuring high capacity, ultra-high reliability, and low latency over multiple simultaneous connections in the same communication network infrastructure not only requires development of 5G/6G mobile fronthaul networks but also requires major developments at the wired network backhaul side.

In order to increase the data transport rate and reduce the signal latency, reconfigurable add/drop multiplexer (ROADM) systems [6,7] are widely deployed in DWDM network nodes, as illustrated in Figure 1a. ROADM systems can facilitate the addition of new services without requiring an expensive upgrade or substantial change to telecom networks [8]. A ROADM allows remote, precise, and flexible selection of wavelengths, so it significantly increases the network capacity without major expense. The ROADM market is forecasted to have tremendous growth following the progress of deployment of 5G/6G networks.

Referring to Figure 1b, the $1 \times N$ wavelength selective switch (WSS) and $M \times N$ Add/Drop WSS (AdWSS) are two core subsystems of a ROADM system [9]. The optical system of a WSS, as shown in Figure 2a, performs wavelength de-multiplexing, beam steering, and then wavelength multiplexing. A WSS system can direct an optical beam at any wavelength to any desired output port. Several technologies have been used for light beam steering, such as micro-electromechanical system (MEMS) arrays [10], liquid crystal cells and crystal wedges [11], and liquid-crystal-on-silicon (LCoS) phase modulators [12]. Among them, LCoS phase modulators are used in most current WSS systems because they are easier to develop high port count WSS systems and to realize dynamic control of the channel center with high resolution [13,14]. As Figure 2b depicts, a blazed grating is formed on the LCoS phase modulator corresponding to each incident beam (e.g., red, green, and blue) when a designed profile of spatially dependent voltages is applied to the pixels. When a light beam is incident upon the grating, the reflected beam is diffracted to a designed direction. Changing the blazed grating pitch enables the grating to diffract an incident optical beam to different angles, resulting in $1 \times N$ optical switching.

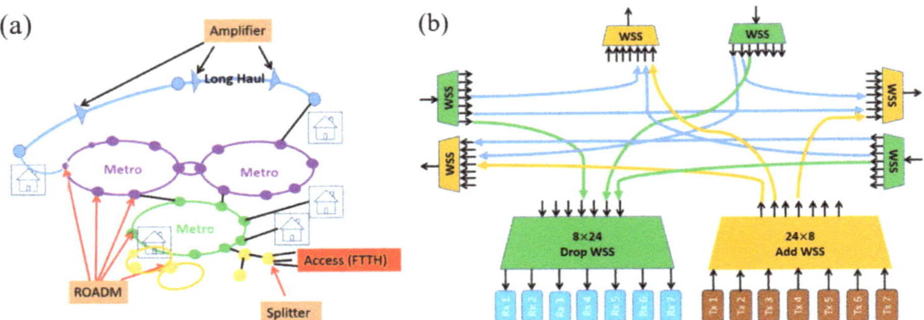

Figure 1. (**a**) Schematic diagram of optical telecommunication network; (**b**) Example of ROADM architecture.

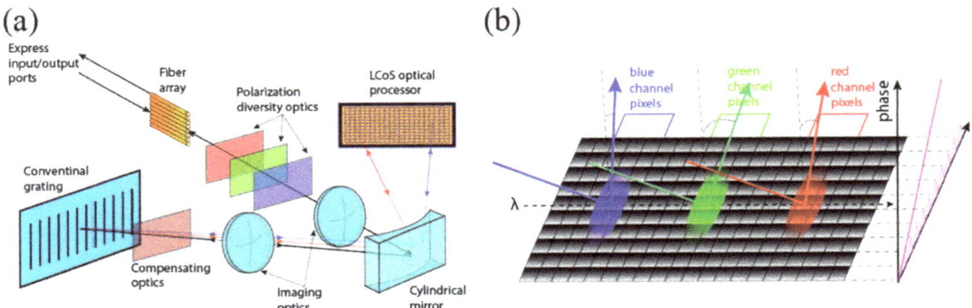

Figure 2. (**a**) The optical system of WSS; (**b**) illustration of beam steering principle using an LCoS phase modulator.

Currently, the optical network industry faces significant challenges to develop high port count, high switching speed, and low insertion loss WSS systems. Generally, current WSS systems have a beam steering time longer than 1 s. Such a steering time consists of two parts: the frame data loading time onto LCoS silicon backplane and the liquid crystal (LC) response time [15]. The latter is usually much slower than the former. For example, the frame data refreshing time is ~8 ms when an LCoS phase modulator is operated at a frame rate of 120 Hz. The frame data loading time can be further reduced by increasing the frame rate if necessary. In current WSS systems, LC response time is much longer than data loading time, resulting in a slow beam steering speed. Therefore, to increase the beam steering speed, we need to develop fast-response LC materials.

In a WSS, to achieve the required hitless switching, beam steering is performed with several steps, for example, 4–6 steps. Therefore, increasing the LC switching speed can significantly reduce the system's beam steering time. For example, if the LC switching time is less than 10 ms, then the system's beam steering time can be reduced by one order from current millisecond switching time. More details about how LC materials influence the performance of WSS can be found in Ref. [16].

Fast-response LC materials have been developed for display applications [17] and phase-only modulators for beam steering applications [15,18,19]. Most commercially available LC mixtures exhibit a birefringence $\Delta n < 0.3$ at $\lambda = 589.3$ nm in order to keep a good photostability [16]. However, for optical telecommunications, the wavelength is typically at $\lambda = 1550$ nm. As the wavelength increases, Δn gradually decreases and then saturates [15]. Moreover, for an LCoS phase modulator, the required phase change is 2π, instead of 1π for the corresponding intensity modulator. That means that, if we use the same LC material for intensity and phase modulators, the latter will have a 4× slower response time. Thus, it is really challenging to achieve a fast response time for the intended 5G/6G communications.

In this paper, we evaluate two fast-response LC materials, LCM-1107 and LCM-2018 (LC Matter). Both mixtures exhibit a high Δn and relatively low viscosity. The measured response time is less than 10 ms at 5 V, $\lambda = 1550$ nm, and the intended operating temperature of an LCoS, which is between 40 and 60 °C. A distributed Bragg reflector is also designed to protect these high Δn LC materials from being damaged by the ambient UV (ultraviolet) and blue lights.

2. Material Characterizations

The physical properties of LCM-1107 and LCM-2018 are listed in Table 1. Differential scanning calorimetry (DSC, TA Instruments Q100) was applied to measure the melting temperature (T_m) and clearing temperature (T_c). From Table 1, these two materials show a wide nematic range to meet the requirements for LCoS applications in 6G communications. The dielectric constants were measured with a multifrequency LCR meter, HP-4274. Both mixtures have a reasonably large dielectric constant ($\Delta\varepsilon > 15$), which helps to lower the threshold voltage. According to the measured free relaxation response time of the test cell [20], the viscoelastic constant γ_1/K_{11} of each mixture can be extracted. For comparison, we also include the properties of a Merck high birefringence LC material, BL038 [21], as shown in Table 1. From Table 1, we can see that these two materials exhibit high Δn, large dielectric constants, and relatively low viscoelastic constants.

Table 1. Measured physical properties of LCM-1107, LCM-2018, and BL038 at $T = 22$ °C.

LC Mixture	LCM-1107	LCM-2018	BL038
T_c (°C)	99.2	115.6	100.0
T_m (°C)	<−20	<−20	-
Δn@1550 nm	0.312	0.344	0.257(@633 nm)
$\Delta\varepsilon$@1 kHz	16.3	17.3	14.4
$\varepsilon_\perp$@1 kHz	4.84	5.07	5.0
K_{11} (pN)	12.7	14.6	19.1
γ_1/K_{11} (ms/μm^2)	17.6	14.5	30.2

2.1. Birefringence

To measure Δn, LCM-1107 and LCM-2018 were filled into two homogeneous cells, whose cell gaps are $d = 8.03$ μm and $d = 8.10$ μm, respectively. The pretilt angle of the rubbed polyimide alignment layers is about 3°. Then, the cell was mounted on a Linkam heating stage, and the temperature was controlled through a temperature programmer TMS94. The test cell was sandwiched between two crossed polarizers and the LC director was oriented at 45° with respect to the optical axis of the polarizer. During the measurement, the sample was activated with a 1 kHz square-wave AC voltage, and the input wavelength ($\lambda = 1550$ nm) was from a laser diode. The transmitted light was detected by an infrared (IR) detector, so that we could obtain the voltage–transmittance (V–T) curve. This V–T curve can be converted to a voltage–phase (V–Φ) curve [22].

Next, the birefringence at a temperature can be calculated from the measured phase retardation. Figure 3 shows the measured (dots and triangles) and fitting (solid line) results of these two materials at different temperatures. The temperature-dependent birefringence can be described by the following equation [23]:

$$\Delta n = \Delta n_0 S = \Delta n_0 (1 - T/T_c)^{\beta}. \quad (1)$$

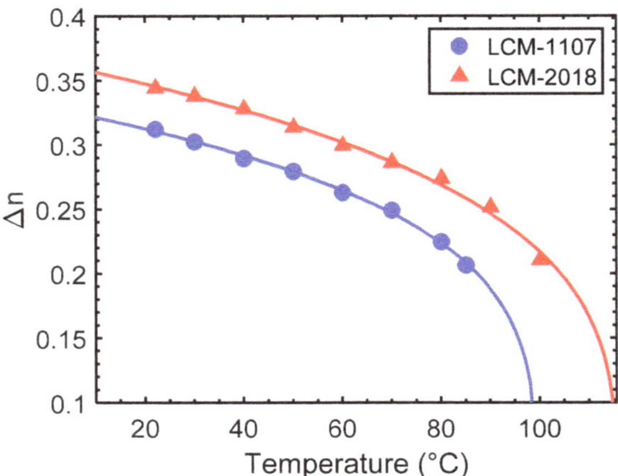

Figure 3. Temperature-dependent birefringence of LCM-1107 and LCM-2018 at λ = 1550 nm and f = 1 kHz. Dots and triangles are measured data and solid lines are fitting curves with Equation (1).

In Equation (1), S is the order parameter, T_c (unit: K) is the clearing temperature of the LC (measured by DSC), Δn_0 represents the extrapolated birefringence when T = 0 K, and exponent β is a material parameter. The values of both Δn_0 and β can be obtained by fitting the experimental data with Equation (1), whose results are listed in Table 2.

Table 2. Fitting parameters obtained through Equations (1)–(3).

LC Mixture	Δn_0	β	G@40 °C (μm^{-2})	λ*@40 °C (μm)	A (ms/μm^2)	E_a (meV)
LCM-1107	0.451	0.236	3.67	0.276	1.69×10^{-5}	342
LCM-2018	0.500	0.260	3.91	0.284	1.16×10^{-4}	288

For LiDAR (light detection and ranging) and 6G optical communications, the employed wavelength could be different. Therefore, we also measured the wavelength-dependent birefringence of these two materials. In the experiment, we measured the wavelength dispersion at the working temperature of LCoS, which is typically 40–60 °C. Six different wavelengths were used: a He-Ne laser at λ = 632.8 nm, a tunable Argon ion laser with λ = 457, 488, and 514 nm, and two diode IR lasers at λ = 1060 and 1550 nm. The measured results (dots and triangles) are plotted in Figure 4. We can fit the experimental results with the following single-band birefringence dispersion equation [23]:

$$\Delta n = G \frac{\lambda^2 \lambda^{*2}}{\lambda^2 - \lambda^{*2}}. \tag{2}$$

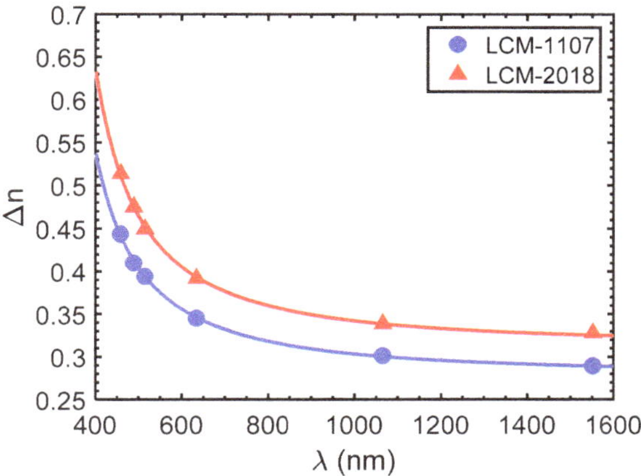

Figure 4. Wavelength-dependent birefringence of LCM-1107 and LCM-2018 at 40 °C and 1 kHz. Dots and triangles are measured data and solid lines are fitting curves with Equation (2).

Equation (2) has two fitting parameters: G (proportionality constant) and λ^* (mean resonance wavelength). The fitting results are listed in Table 2.

2.2. Viscoelastic Constant

The viscoelastic constant γ_1/K_{11} is also an important parameter affecting the LCoS response time. We measured γ_1/K_{11} through the transient decay time of an LC cell. The obtained results are plotted in Figure 5, where dots and triangles are the measured results, and solid lines are the fitting results using the following equation [24]:

$$\frac{\gamma_1}{K_{11}} = A\frac{\exp(E_a/k_B T)}{(1 - T/T_c)^{\beta}}. \tag{3}$$

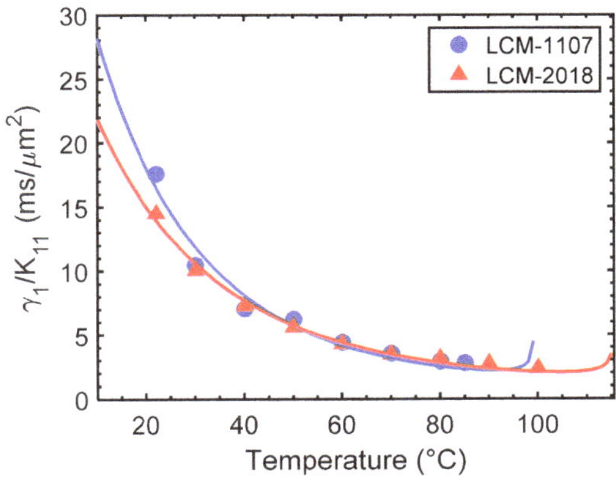

Figure 5. Temperature-dependent viscoelastic constant of LCM-1107 and LCM-2018. Dots and triangles are measured data and solid lines are fitting curves with Equation (3).

In Equation (3), A is the proportionality constant, E_a is the activation energy, and k_B is the Boltzmann constant. Here, A and E_a are fitting parameters, whose values are also included in Table 2.

2.3. Figure of Merit

Figure of merit, which is defined as $FoM = \Delta n^2/(\gamma_1/K_{11})$, is a fair way to compare the electro-optic performance of different materials [24]. Based on the results in Figures 3 and 5, we can plot the temperature-dependent FoM as shown in Figure 6, where dots are experimental data and solid lines are the fitting curves using the definition of *FoM*, and Equation (2) for Δn and Equation (3) for γ_1/K_{11}. From the figure, we can see that as the temperature increases, FoM increases first, reaching a peak, and then declines sharply as the temperature approaches T_c. The reason the FOM shows such a tendency is that both Δn and γ_1/K_{11} decrease but at different rates as the temperature initially increases. In Figure 3, Δn decreases slowly as the temperature gradually increases, but the decreasing rate becomes more pronounced as the temperature approaches T_c. However, γ_1/K_{11} shows an opposite trend, i.e., it decreases rapidly in the beginning and then gradually saturates as the temperature gets closer to T_c. Detailed mechanisms have been explained in Ref. [24].

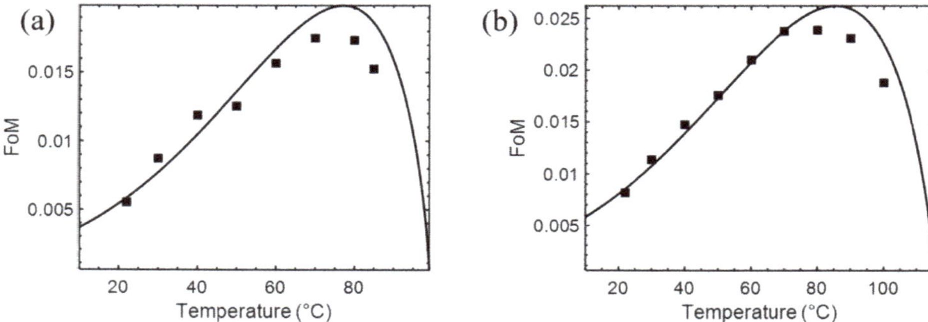

Figure 6. Temperature-dependent FoM of (**a**) LCM-1107 and (**b**) LCM-2018. Squares are measured data and solid lines are fitting curves.

2.4. Voltage-Dependent Phase Change

For a phase-only LCoS, we can select an optimal cell gap for each LC material to achieve 2π phase change at the maximally allowable operation voltage, which is 5 V ($V_{2\pi}$ = 5 V), and the intended operation temperature and wavelength. In the experiment, for convenience we measured the voltage-dependent transmittance (V–T) curves of these two materials using transmission-type LC cells at λ = 1550 nm and then converted to the voltage-dependent phase (V–Φ) curves depicted by red lines in Figure 7. Since the operating temperature of a working LCoS device could vary from 40 to 60 °C due to the thermal effects from the backplane driving circuits and the employed light source, we measured the electro-optic effects at 40, 50, and 60 °C for LCM-1107 (Figure 7a–c) and LCM-2018 (Figure 7d–f), respectively. From the measured results (red lines) shown in Figure 7, when the applied voltage is 5 V, the phase change is more than 2π, which means we can use a thinner cell gap (optimal cell gap) to achieve the desired $V_{2\pi}$ = 5 V at each temperature. Due to the limited available cell gaps in our lab, we extrapolated the V–Φ curve to the corresponding optimal cell gap as shown by the blue lines in Figure 7. The optimal cell gaps and the extrapolated V–Φ curves are obtained according to the phase retardation equation $\delta = 2\pi d\Delta n/\lambda$. Since the cell gap is thinner, the response time is faster, as will be discussed quantitatively later.

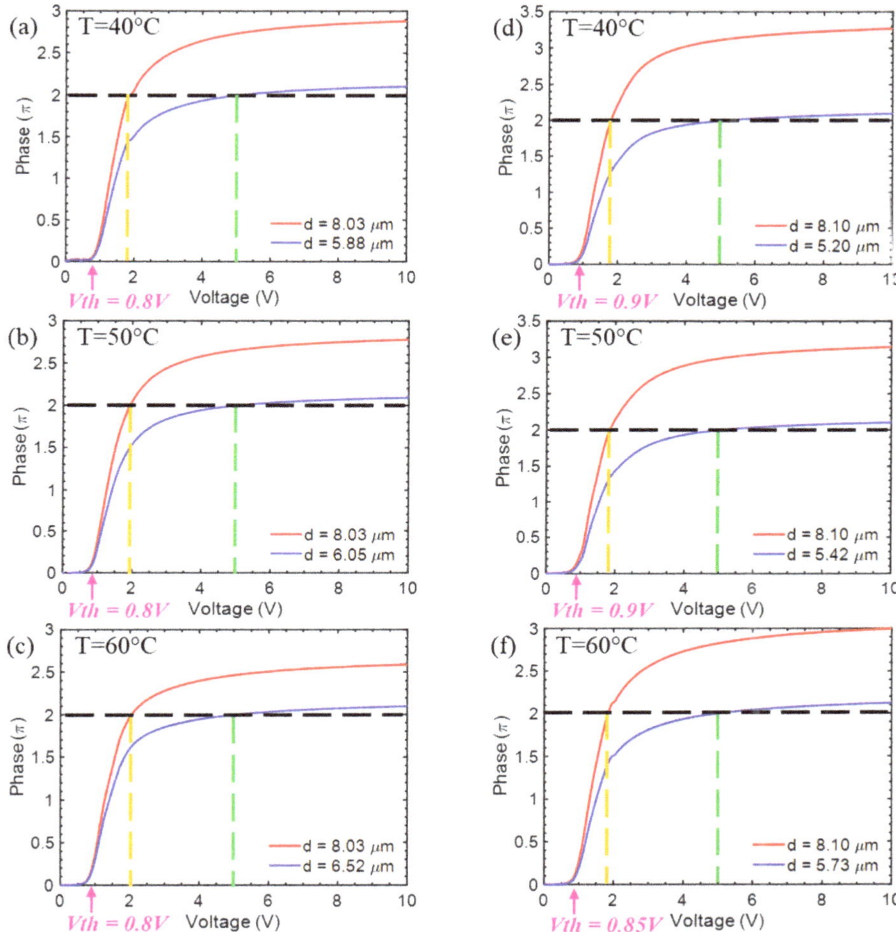

Figure 7. Measured (red lines) voltage-dependent phase change in a transmissive LCM-1107 cell (**a–c**) and LCM-2018 cell (**d–f**) at λ = 1550 nm, and the extrapolated (blue lines) voltage-dependent phase change of the corresponding optimal cell gaps at $V_{2\pi}$ = 5 V_{rms}.

2.5. Response Time

The response time (rise time and decay time) of an LC phase modulator between two gray levels (V_1 and V_2) depends on the viscoelastic constant, LC cell gap, threshold voltage, and operating voltages, which are described in the following equations [25]:

$$\tau_{on} = \frac{\tau_0}{(V_2/V_{th})^2 - 1}, \tag{4}$$

$$\tau_{off} = \frac{\tau_0}{\left|(V_1/V_{th})^2 - 1\right|}, \tag{5}$$

$$\tau_0 = \frac{\gamma_1 d^2}{K_{11}\pi^2}. \tag{6}$$

In Equations (4)–(6), τ_0 is the free relaxation time, V_{th} is the threshold voltage, V_1 and V_2 are the low and high gray-level voltages, respectively. Based on the previous studies [17,26], the average gray-to-gray rise and decay time is approximately equal to the sum of free relaxation time ($V_1 = 0$) and turn-on time (V_2 = 8th gray level). Therefore, in our experiment we set $V_1 = 0$ and $V_2 = V_{2\pi}$ and just measured the rise time and decay time between gray levels 1 and 8 for each test cell without the need of measuring all the gray-to-gray transition times.

In experiment, the rise time and decay time of LCM-1107 and LCM-2018 were measured at T = 40, 50 and 60 °C at $V_{2\pi}$ = 5 V and λ = 1550 nm. Results are shown in Table 3 for LCM-1107 and Table 4 for LCM-2018. Next, we converted the measured results to the corresponding reflection-type cell, which is four times faster than that of a transmissive cell because its cell gap is one-half of the transmissive one. Furthermore, in each scenario we also need to convert the response time to the corresponding optimal cell gap with $V_{2\pi}$ = 5 V, as the blue lines depict in Figure 7. Under optimal conditions, the cell gap is thinner and the value of $V_{2\pi}$ increases to 5 V, which contributes to a faster rise time, as Equation (4) indicates. On the other hand, a thinner cell gap also helps to improve the decay time according to Equations (5) and (6). Therefore, the total response time is reduced significantly by using an optimal cell gap, and the extrapolated results are shown in Table 3 for LCM-1107 and Table 4 for LCM-2018. From Table 3, LCM-1107 has a total response time of 6 ms at 50 °C. On the other hand, LCM-2018 has an even higher birefringence than LCM-1107. Thus, a thinner cell gap can be used, resulting in a 5.1 ms total response time at 50 °C and 4.6 ms one at 60 °C. That means that by using LCM-2018 we can achieve a ~200 Hz frame rate for 6G communication at λ = 1550 nm.

Table 3. Measured response time of a transmissive LCM-1107 cell with d = 8.03 μm and the extrapolated response time to the corresponding reflective cells at 40, 50 and 60 °C with λ = 1550 nm.

T (°C)	d (μm)	V_{th} (V)	$V_{2\pi}$ (V)	τ_{on} (ms)	τ_{off} (ms)	τ_{total} (ms) Transmissive	τ_{total} (ms) Reflective
40	8.03	0.8	1.88	40.4	56.6	97.1	24.3
40	5.88	0.8	5.0	2.4	30.4	32.8	8.2
50	8.03	0.8	1.94	27.1	39.2	66.3	16.6
50	6.05	0.8	5.0	1.9	22.3	24.2	6.0
60	8.03	0.8	2.04	17.9	30.9	48.8	12.2
60	6.52	0.8	5.0	1.6	20.4	22.0	5.5

Table 4. Measured response time of a transmissive LCM-2018 cell with d = 8.10 μm and the extrapolated response time to the corresponding reflective cells at 40, 50 and 60 °C with λ = 1550 nm.

T (°C)	d (μm)	V_{th} (V)	$V_{2\pi}$ (V)	τ_{on} (ms)	τ_{off} (ms)	τ_{total} (ms) Transmissive	τ_{total} (ms) Reflective
40	8.10	0.9	1.80	54.4	52.1	106.4	26.6
40	5.20	0.9	5.0	2.3	21.5	23.7	5.9
50	8.10	0.9	1.84	38.3	41.5	79.8	19.9
50	5.42	0.9	5.0	1.8	18.6	20.4	5.1
60	8.10	0.85	1.85	29.8	33.7	63.4	15.9
60	5.73	0.85	5.0	1.7	16.8	18.5	4.6

It should be mentioned here that, in addition to the material parameters and cell gap shown in Equations (4)–(6), the LC response time also depends on the anchoring energy of the alignment layers [27]. Equations (4)–(6) are derived based on strong anchoring energy, such as buffed polyimide alignment layers. If an inorganic SiO_2 layer is used, its anchoring energy could be lower. As a result, the corresponding threshold voltage would decrease, and decay time would increase slightly due to the weaker restoring force [28].

3. Coating Design

Both LCM-1107 and 2018 contain some isothiocyanato (NCS) phenyl-tolane compounds [29,30]. Thus, they exhibit a high Δn, large $\Delta\varepsilon$, and relatively low viscosity. However, due to their long molecular conjugation, the photostability of these high birefringence NCS compounds are usually inadequate, especially in the UV and blue spectral regions. To seal the filling hole of the LCoS panel, we should block the active LC area and use a longer UV wavelength, such as 385 nm, to cure the sealant glue, as discussed in [17].

To protect the high birefringence LC materials from ambient UV and blue light irradiation, we designed a distributed Bragg reflector (DBR) to filter out the wavelength from 300 to 500 nm, while keeping a high transmittance at 1550 nm. Such a DBR film should only cover the active LC area, except for the filling hole, because it is not transparent at 385 nm.

In our design, we applied TiO_2 and SiO_2 as the high and low refractive index materials, respectively, which are commonly used in DBR fabrication [31–33]. The wavelength-dependent refractive indices of TiO_2 and SiO_2 were taken from [31] and [32], respectively. Since we require a broad reflection band in the UV–blue region, in our design we adopted two groups of DBR stacks [33] corresponding to the two reflection peaks at 360 and 480 nm, respectively. Each group consists of eight pairs of TiO_2/SiO_2. Each pair in the first group (corresponding to the 360 nm reflection peak) has a thickness of TiO_2(32 nm)/SiO_2(61 nm), and the second group (corresponding to the 480 nm reflection peak) has the thickness of TiO_2(47 nm)/SiO_2(81 nm). The transmission spectrum of this DBR is shown in Figure 8. The transmittance is less than 0.5% from 300 to 510 nm but is as high as 97.2% at 1550 nm.

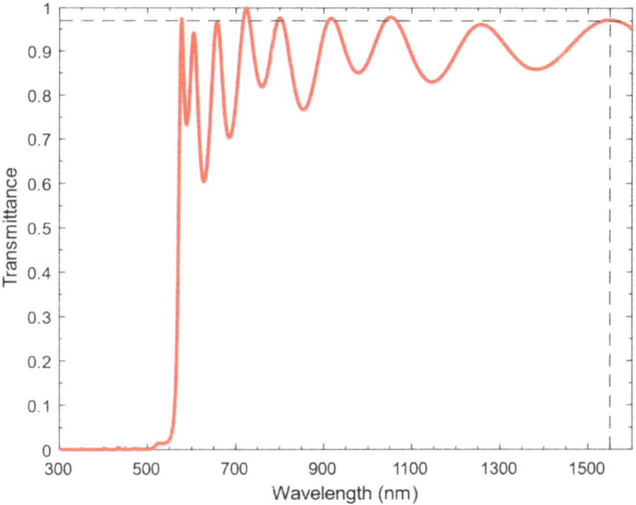

Figure 8. Transmission spectrum of the designed distributed Bragg reflector.

4. Conclusions

We reported two high birefringence LC mixtures, LCM-1107 and LCM-2018, for phase-only LCoS panels intended for 6G communications, whose working wavelength is around 1550 nm. These two materials can easily achieve a total response time of 6 ms at 50 °C operating temperature, while keeping the 2π phase change voltage at $5V_{rms}$. To enhance the photostability, we also designed a DBR coating, which has a transmittance lower than 0.5% from 300 to 510 nm, and higher than 97% at 1550 nm. These two materials are promising candidates for LCoS-based 6G communications.

Author Contributions: Methodology, J.Z. and Z.Y.; writing—original draft preparation, J.Z. and C.M.; writing—review and editing, S.-T.W.; supervision, S.-T.W. All authors have read and agreed to the published version of the manuscript.

Funding: Sony Corporation.

Acknowledgments: The UCF group is indebted to Sony Corp. for the financial support and Sebastian Gauza of LC Matter for providing the liquid crystal materials.

Conflicts of Interest: The authors declare no conflict of interest.

References

1. Brackett, C.A. Dense wavelength division multiplexing networks: Principles and applications. *IEEE J. Sel. Areas Commun.* **1990**, *8*, 948–964. [CrossRef]
2. Zhu, L.; Zhu, G.; Wang, A.; Wang, L.; Ai, J.; Chen, S.; Du, C.; Liu, J.; Yu, S.; Wang, J. 18 km low-crosstalk OAM+ WDM transmission with 224 individual channels enabled by a ring-core fiber with large high-order mode group separation. *Opt. Lett.* **2018**, *43*, 1890–1893. [CrossRef] [PubMed]
3. Panwar, N.; Sharma, S.; Singh, A.K. A survey on 5G: The next generation of mobile communication. *Phys. Commun.* **2016**, *18*, 64–84. [CrossRef]
4. Giordani, M.; Polese, M.; Mezzavilla, M.; Rangan, S.; Zorzi, M. Toward 6G networks: Use cases and technologies. *IEEE Commun. Mag.* **2020**, *58*, 55–61. [CrossRef]
5. Chowdhury, M.Z.; Shahjalal, M.; Ahmed, S.; Jang, Y.M. 6G wireless communication systems: Applications, requirements, technologies, challenges, and research directions. *IEEE Open J. Commun. Soc.* **2020**, *1*, 957–975. [CrossRef]
6. Tibuleac, S.; Filer, M. Transmission impairments in DWDM networks with reconfigurable optical add-drop multiplexers. *J. Lightwave Technol.* **2010**, *28*, 557–568. [CrossRef]
7. Wang, S.; Feng, X.; Gao, S.; Shi, Y.; Dai, T.; Yu, H.; Tsang, H.K.; Dai, D. On-chip reconfigurable optical add-drop multiplexer for hybrid wavelength/mode-division-multiplexing systems. *Opt. Lett.* **2017**, *42*, 2802–2805. [CrossRef]
8. Geng, M.; Jia, L.; Zhang, L.; Yang, L.; Chen, P.; Wang, T.; Liu, Y. Four-channel reconfigurable optical add-drop multiplexer based on photonic wire waveguide. *Opt. Express* **2009**, *17*, 5502–5516. [CrossRef]
9. Ma, Y.; Stewart, L.; Armstrong, J.; Clarke, I.; Baxter, G.W. Recent Progress of Wavelength Selective Switch. *J. Lightwave Technol.* **2020**, *39*, 896–903. [CrossRef]
10. Marom, D.M.; Neilson, D.T.; Greywall, D.S.; Pai, C.S.; Basavanhally, N.R.; Aksyuk, V.A.; López, D.O.; Pardo, F.; Simon, M.E.; Low, Y.; et al. Wavelength-selective 1 x K switches using free-space optics and MEMS micromirrors: Theory, design, and implementation. *J. Lightwave Technol.* **2005**, *23*, 1620–1630. [CrossRef]
11. Scherger, B.; Reuter, M.; Scheller, M.; Altmann, K.; Vieweg, N.; Dabrowski, R.; Deibel, J.A.; Koch, M. Discrete terahertz beam steering with an electrically controlled liquid crystal device. *J. InfraredMillim. Terahertz Waves* **2012**, *33*, 1117–1122. [CrossRef]
12. Baxter, G.; Frisken, S.; Abakoumov, D.; Zhou, H.; Clarke, I.; Bartos, A.; Poole, S. Highly programmable wavelength selective switch based on liquid crystal on silicon switching elements. In Proceedings of the 2006 Optical Fiber Communication Conference and the National Fiber Optic Engineers Conference, Anaheim, CA, USA, 5–10 March 2006.
13. Wang, M.; Zong, L.; Mao, L.; Marquez, A.; Ye, Y.; Zhao, H.; Vaquero, C.F.J. LCoS SLM study and its application in wavelength selective switch. *Photonics* **2017**, *4*, 22. [CrossRef]
14. Frisken, S.; Baxter, G.; Abakoumov, D.; Zhou, H.; Clarke, I.; Poole, S. Flexible and grid-less wavelength selective switch using LCOS technology. In Proceedings of the 2011 Optical Fiber Communication Conference and Exposition and the National Fiber Optic Engineers Conference, Los Angeles, CA, USA, 6–10 March 2011.
15. Zou, J.; Yang, Q.; Hsiang, E.L.; Ooishi, H.; Yang, Z.; Yoshidaya, K.; Wu, S.T. Fast-response liquid crystal for spatial light modulator and LiDAR applications. *Crystals* **2021**, *11*, 93. [CrossRef]
16. Lazarev, G.; Chen, P.J.; Strauss, J.; Fontaine, N.; Forbes, A. Beyond the display: Phase-only liquid crystal on Silicon devices and their applications in photonics. *Opt. Express* **2019**, *27*, 16206–16249. [CrossRef]
17. Huang, Y.; He, Z.; Wu, S.T. Fast-response liquid crystal phase modulators for augmented reality displays. *Opt. Express* **2017**, *25*, 32757–32766. [CrossRef]
18. Yang, Q.; Zou, J.; Li, Y.; Wu, S.T. Fast-response liquid crystal phase modulators with an excellent photostability. *Crystals* **2020**, *10*, 765. [CrossRef]
19. He, Z.; Yin, K.; Wu, S.T. Miniature planar telescopes for efficient, wide-angle, high-precision beam steering. *Light Sci. Appl.* **2021**, *10*, 134. [CrossRef] [PubMed]
20. Wu, S.T.; Wu, C.S. Rotational viscosity of nematic liquid crystals a critical examination of existing models. *Liq. Cryst.* **1990**, *8*, 171–182. [CrossRef]
21. Dąbrowski, R.; Kula, P.; Herman, J. High Birefringence Liquid Crystals. *Crystals* **2013**, *3*, 443–482. [CrossRef]
22. Wu, S.T.; Efron, U.; Hess, L.D. Birefringence measurements of liquid crystals. *Appl. Opt.* **1984**, *23*, 3911–3915. [CrossRef] [PubMed]
23. Wu, S.T. Birefringence dispersions of liquid crystals. *Phys. Rev. A* **1986**, *33*, 1270–1274. [CrossRef]

24. Wu, S.T.; Lackner, A.M.; Efron, U. Optimal operation temperature of liquid crystal modulators. *Appl. Opt.* **1987**, *26*, 3441–3445. [CrossRef]
25. Wu, S.T. Design of a liquid-crystal-based electro-optic filter. *Appl. Opt.* **1989**, *28*, 48–52. [CrossRef]
26. Chen, H.; Gou, F.; Wu, S.T. Submillisecond-response nematic liquid crystals for augmented reality displays. *Opt. Mater. Express* **2017**, *7*, 195–201. [CrossRef]
27. Nie, X.; Xianyu, H.; Lu, R.; Wu, T.X.; Wu, S.-T. Anchoring energy and cell gap effects on liquid crystal response time. *J. Appl. Phys.* **2007**, *101*, 103110. [CrossRef]
28. Jiao, M.; Ge, Z.; Song, Q.; Wu, S.T. Alignment layer effects on thin liquid crystal cells. *Appl. Phys. Lett.* **2008**, *92*, 061102. [CrossRef]
29. Gauza, S.; Li, L.; Wu, S.T.; Spadlo, A.; Dabrowski, R.; Tzeng, Y.N.; Cheng, K.L. High birefringence and high resistivity isothiocyanate-based nematic liquid crystal mixtures. *Liq. Cryst.* **2005**, *32*, 1077–1085. [CrossRef]
30. Gauza, S.; Wang, H.; Wen, C.H.; Wu, S.T.; Seed, A.; Dabrowski, R. High birefringence isothiocyanato tolane liquid crystals. *Jpn. J. Appl. Phys.* **2003**, *42*, 3463–3466. [CrossRef]
31. Lin, K.C.; Lee, W.K.; Wang, B.K.; Lin, Y.H.; Chen, H.H.; Song, Y.H.; Huang, Y.H.; Shih, L.W.; Wu, C.C. Modified distributed Bragg reflector for protecting organic light-emitting diode displays against ultraviolet light. *Op. Express* **2021**, *29*, 7654–7665. [CrossRef] [PubMed]
32. Gao, L.; Lemarchand, F.; Lequime, M. Refractive index determination of SiO_2 layer in the UV/Vis/NIR range: Spectrophotometric reverse engineering on single and bi-layer designs. *J. Eur. Opt. Soc. Rapid Publ.* **2013**, *8*, 13010. [CrossRef]
33. Ding, X.; Gui, C.; Hu, H.; Liu, M.; Liu, X.; Lv, J.; Zhou, S. Reflectance bandwidth and efficiency improvement of light-emitting diodes with double-distributed Bragg reflector. *Appl. Opt.* **2017**, *56*, 4375–4380. [CrossRef] [PubMed]

Article

Identifying a Unique Communication Mechanism of Thermochromic Liquid Crystal Printing Ink

Maja Strižić Jakovljević [1,*], Branka Lozo [1] and Marta Klanjšek Gunde [2,*]

[1] Faculty of Graphic Arts, University of Zagreb, Getaldićeva 2, 10000 Zagreb, Croatia; branka.lozo@grf.hr
[2] National Institute of Chemistry, Hajdrihova 19, 1001 Ljubljana, Slovenia
* Correspondence: maja.strizic.jakovljevic@grf.unizg.hr (M.S.J.); marta.k.gunde@ki.si (M.K.G.); Tel.: +385-2371-080 (ext. 255) (M.S.J.); +386-147-60-291 (M.K.G.)

Abstract: Thermochromic liquid crystal materials are commonly used in printing inks, opening up a wide range of possible applications. In order to ensure and control the most accurate application, the occurrence of the so-called colour play effect, i.e., the appearance of iridescent (rainbow) colours as a function of temperature, must be determined precisely. For this purpose, the temperature-dependent reflection of a sample must be measured using a spectrometer with an integrating sphere. The same values should be obtained for each sample containing the same thermochromic liquid crystalline material, irrespective of the spectrometer used, integrating sphere, layer thickness and the surface properties of the substrate. To describe this intrinsic property of the thermochromic liquid crystal material, the term communication mechanism might be considered. The research has shown how this mechanism is obtained experimentally.

Keywords: thermochromic liquid crystal inks; temperature; colour play effect; communication mechanism

Citation: Strižić Jakovljević, M.; Lozo, B.; Gunde, M.K. Identifying a Unique Communication Mechanism of Thermochromic Liquid Crystal Printing Ink. *Crystals* **2021**, *11*, 876. https://doi.org/10.3390/cryst11080876

Academic Editors: Kohki Takatoh, Jun Xu and Akihiko Mochizuki

Received: 13 July 2021
Accepted: 27 July 2021
Published: 28 July 2021

1. Introduction

Thermochromic liquid crystal (TLC) inks respond to temperature change with a change of colour [1]. In order to achieve numerous possible applications, the TLC functional material is usually microencapsulated, so as to protect its unique properties and to "pigment" the ink or some other host material. The TLC material inside the microcapsules determines the colour, mechanism of colour change, and temperature at which the change occurs, but the binder of the ink defines its printing and curing technology [2,3].

TLC inks are coloured within the temperature activation range of several degrees, also referred to as the "bandwidth" or "colour play interval" [4,5]. At temperatures below or above the activation range, the TLC ink is colourless. During heating, at the point of reaching the activation temperature, the red colour appears first, followed by orange, yellow, green, blue, and violet. This effect is referred to as "colour play" [4,6]. Each of the colours is limited to a narrow temperature interval [3]. Above the upper threshold of the activation range, the violet colour disappears and the TLC ink becomes colourless again. The temperature required to reach the colourless stage is called the "clearing point" [4,7,8]. Our previous experiments have shown the colour cycles of TLC inks to be reversible.

Microcapsules are the functional pigments and can be used in a number of different applications. They are usually contained in a binder system and are incorporated in various places including wearable devices [9–11]. In this study, we used the TLC inks in which the binder of the ink defines the printing and curing technology, whereas the pigment defines the thermochromic functionality [2,3]. Such inks can be used to print arbitrary designs on various surfaces, including direct printing on curved ones. In this special state, the adjacent sheets of equally oriented molecules twist and the corresponding director (i.e., the direction of the long axis of the molecules) traces a helical path. The distance required for the director to complete a 360° turn is called the pitch length. The thermochromic effect of

TLCs results from the temperature-dependent pitch. With an increase in temperature, the helical pitch shrinks, leading to reflections of light with shorter wavelengths [12,13].

The "colour play effect" of TLC inks is only clearly visible when the ink is printed on a black substrate [7,14]. The reflection of light from the helical structure is nearly imperceptible, because most of the light passes through the ink layer and hits the substrate. On a white substrate, most of this light is backscattered, virtually obscuring the low light intensity reflected from the molecular pitch [7]. To prevent this obscuring effect, TLCs should be deposited on a black substrate, which can absorb the greatest part of the light transmitted through the ink layer. Under such circumstances, the weak spectral reflection prevails, making the iridescent colours clearly visible [15,16]. Our previous studies have shown that the colour play effect of TLC ink can be observed when it is printed on a grey substrate with an optical density of at least 0.72 [15].

TLC inks are best known as temperature indicators, especially for packaging, security printing and brand protection [2,4,12]. In electronics, liquid crystals can be used to detect electrical shorts in circuits, open circuits, and non-functional devices [12]. Temperature changes of TLCs can be a great advantage in monitoring and mapping the temperature of a significant surface area of almost any shape to detect a temperature fault or locate thermal activity [4,5,17,18]. The unique thermochromic effect of TLCs possesses the potential for an increased application in security printing.

The thermoreactive properties of TLC inks must be measured using integrating spheres, which spatially integrate the radiant flux reflected on a sample in each direction. Our previous studies have shown that larger diameter spheres give better results than smaller ones [7].

To date, no detailed optical analysis of the colour play effect has been performed to show how this effect can be assessed independently of measurement devices, layer thickness, properties of the substrate and the differences between samples obtained from different producers. Therefore, the goal of this research was to analyse how these parameters could be experimentally quantified. We have chosen to call this set of parameters the communication mechanism. It is an intrinsic property of the functional TLC material and is fully independent of application and measurement conditions.

2. Materials and Methods

In the research, we used the water-based TLC ink formulation by Printcolor, Switzerland. According to manufacturer's data, the ink activates at 25 °C (T_A, activation temperature) and the clearing point is at 44 °C. At 25 °C the ink turns red, at 26 °C to green, proceeding to blue at 30 °C. Below T_A and above the cleating point, the ink is colourless.

Two types of printing substrates were used: black coated paper (260 g/m^2, BYK, Geretsried, Germany) and black uncoated paper (160 g/m^2, Hahnemühle, Dassel, Germany). Black coated paper has a thickness of 350 µm, while black uncoated paper is 232 µm thick. The TLC ink was screen-printed over the black substrates with single, double and triple layers (wet over dry), using a semiautomatic screen-printing machine. The SEFAR® PET 1500 43/110-80 W polyester mesh (Sefar AG, Heiden, Germany) with 149 µm openings was used [19]. The prints were dried inside a hot air tunnel at about 75 °C.

Temperature-dependent optical properties of TLC prints were measured in a temperature range from 26 to 79 °C, using two different spectrometers, a full-size scientific device, mostly used in basic research and a fibre-based portable one typical for various spectrometric analyses. Both instruments have an integrating sphere measurement cell with (8°:di) measuring geometry. The full-size spectrometer was Lambda 950 UV-VISNIR (Perkin Elmer, Hopkinton, MA, USA) equipped with a 150 mm wide integrating sphere with a 25 mm sampling port diameter. The fibre-based USB 2000+ spectrometer (Ocean Optics, Orlando, FL, USA) has a 50 mm wide integrating sphere (ISP-50-8-R-GT) and an 8 mm sampling port diameter. SpectraSuite software by Ocean Optics was used to calculate the CIELAB L^*, a^*, b^* values taking into account the D50 illuminant and 2° standard observer. At each individual temperature degree, the reflectance spectra of the samples were measured with 1 nm step in the 350–850 nm spectral region.

Temperature control of the printed samples was carried out using the surface of a water block (EK Water Blocks; EKWB d.o.o., Ljubljana, Slovenia). Thermostatically controlled water circulates through very thin acrylic channels inside the base plate. The heat from the water quickly transfers from thin channels through highly polished copper-nickel plate to the sample. The applied thermostatic circulator allows a heating rate of about 3.7 °C min^{-1}, and the water temperature is accurate up to a tenth of a degree [7,20]. Spectrometric measurements were performed at a steady temperature of the samples. This approach follows the uniform surface temperature method used in liquid crystal thermography [18,20]. Steady temperature of the measured sample is very important to ensure that the temperature of the sample remains constant throughout the sampling of each reflectance spectrum in both measuring devices used.

3. Results and Discussion

The reflectance spectra of the TLC samples prepared on black uncoated paper and measured with the Lambda 950 spectrometer are shown in Figure 1. The spectral characteristic of the colour play effect is a single reflection peak that occurs when the functional material inside the "pigments" is in the chiral nematic/cholesteric phase. This peak moves across the visible range as a function of the temperature. In the example shown, the chiral nematic phase of the liquid crystalline material is formed at 27 °C, where a low and broad reflectance peak appears at the long wavelength limit of the visible (780 nm) (Figure 1). At 28.5 °C, this peak shifts to 713 nm and the sample appears reddish. A further increase in temperature shifts the peak throughout the visible spectrum to shorter wavelengths, making it narrower and more intense. These changes are not linearly dependent on the temperature, so that the blue part of the spectrum remains visible most of the time. Here, the reflectance peak is the narrowest, highly intense, and necessitates more heating to shift to even smaller wavelengths. This corresponds to the visual perception of the sample—its appearance is very distinctly blue and remains as such over the widest temperature range. As the temperature rises above 46 °C, the peak shifts outside of the visible region (below 380 nm) and disappears completely above 75 °C, because the TLC turns into isotropic liquid. This temperature is commonly referred to as the clearing point [7]. It is important to note, however, that the result we obtained is higher than 44 °C as declared by the producer. Therefore, in the studied TLC, both activation and clearing temperatures are outside of the visible spectral range and the activation range is wider than visible. Figure 1 presents the spectra with reflectance peak positioned inside the visible spectral range.

The same sample was also measured with the fibre-optic spectrometer, and the obtained spectra are shown in Figure 2. The results are slightly different from those measured with the Lambda 950 spectrometer (Figure 1), but the basic characteristics of the reflection appear very similar in measurements performed with both instruments.

More commonly applicable results were investigated by colorimetric analysis, as shown in Figure 3. The CIELAB colour values were calculated from the corresponding reflectance spectra measured on the same sample with both spectrometers (Figures 1 and 2). The red-green (a^*) and yellow-blue (b^*) values start and end at almost the same points of the (a^*,b^*) graph, showing the colour of the sample in isotropic phase of the TLC, where no colour is developed. At intermediate temperatures, where the chiral-nematic/cholesteric phase produces the single reflection peak, the entire loop is formed, exhibiting the thermochromic effect. Similar loops were obtained for both measurements. However, some differences do occur, especially in green and blue part of the CIELAB colour space. We assume that these differences are attributable to different measuring equipment used, specifically by the amount of light available in the short wavelength spectral region and the characteristics of the integrating sphere used in both spectrometers. In addition, the fibre-optic spectrometer USB 2000+ (Ocean Optics, Orlando, FL, USA) was used to measure lesser temperature values.

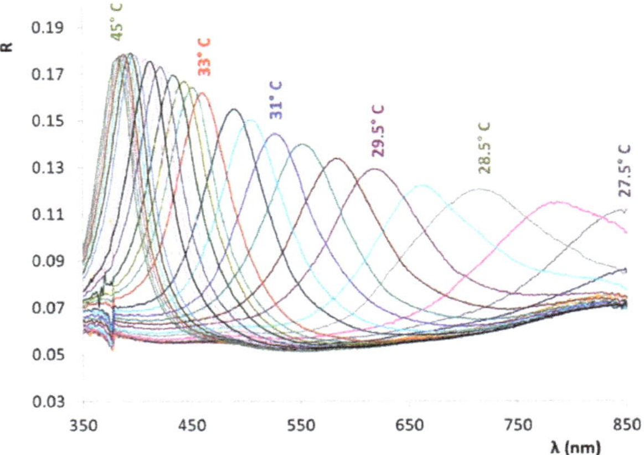

Figure 1. Reflectance spectra of the sample double printed on black uncoated paper, measured in temperature range between 27 and 45 °C with Lambda 950. Only spectra with reflectance peaks in the visible are shown. The temperature of some reflectance spectra is given with the same colour as the corresponding spectrum.

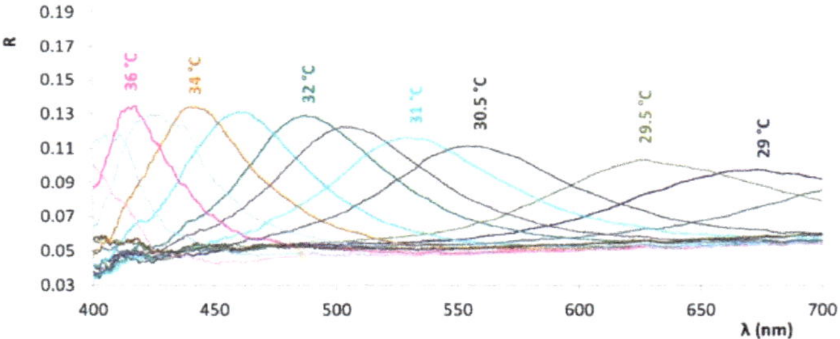

Figure 2. Reflectance spectra of the sample double printed on black uncoated paper, measured in the temperature range between 27.5 and 45 °C with a USB 2000+ spectrometer. The temperature of some reflectance spectra is given with the same colour as the corresponding spectrum.

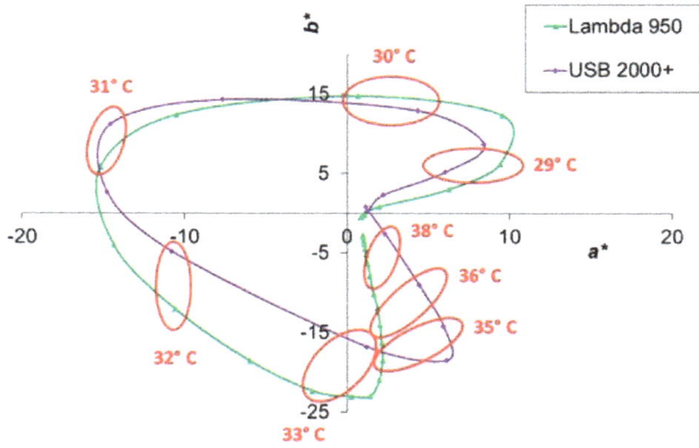

Figure 3. CIELAB colour values of the TLC ink printed on black uncoated paper, measured by both spectrometers and presented in (a^*,b^*) diagram. Red ellipses connect the colour states obtained at the same temperature.

The optical effects of the TLC ink were further analysed as a function of layer thickness. Test samples were prepared by applying single-, double- and triple-printed TLC layers on the coated black paper. The thickness of these samples, as measured by micrometre were 27.6, 34.2 and 46.8 μm for single-, double-, and triple-printed samples, respectively.

The reflectance spectra of these three samples, measured with the Lambda 950 spectrometer, are shown in Figures 4–6. In general, the intensity of the reflection peak is the lowest for the single-printed sample (Figure 4) and the highest for the triple-printed one (Figure 6). More specifically, the highest reflectance peak (0.10) was demonstrated at 45 °C for the single-printed layer, the value of 0.13 appeared in the double-printed one at 39 °C and increased to 0.16 in the triple-printed sample at 41 °C. Thicker layers contain more thermochromic pigments, resulting in a stronger optical effect.

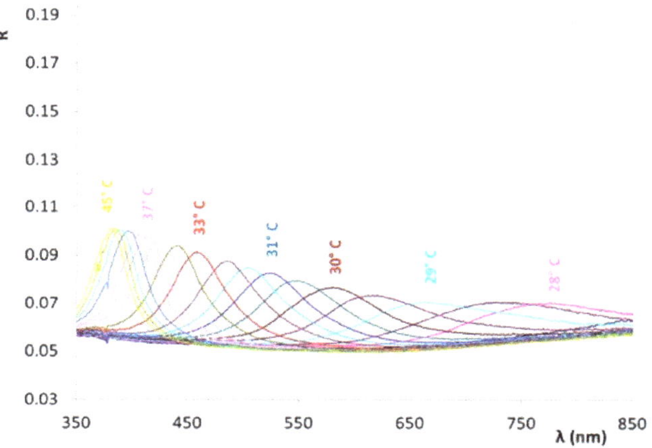

Figure 4. Reflectance spectra of single-printed TLC layer on coated black paper, measured by the Lambda 950 spectrometer. The temperature of some reflectance spectra is given with the same colour as the corresponding spectrum.

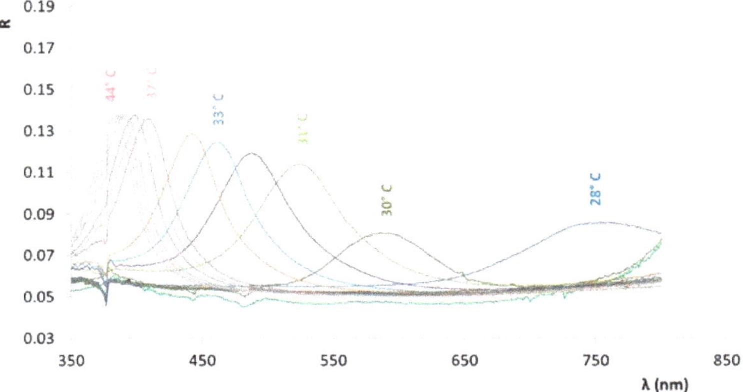

Figure 5. Reflectance spectra of double-printed TLC layer on coated black paper, measured by the Lambda 950 spectrometer. The temperature of some reflectance spectra is given with the same colour as the corresponding spectrum.

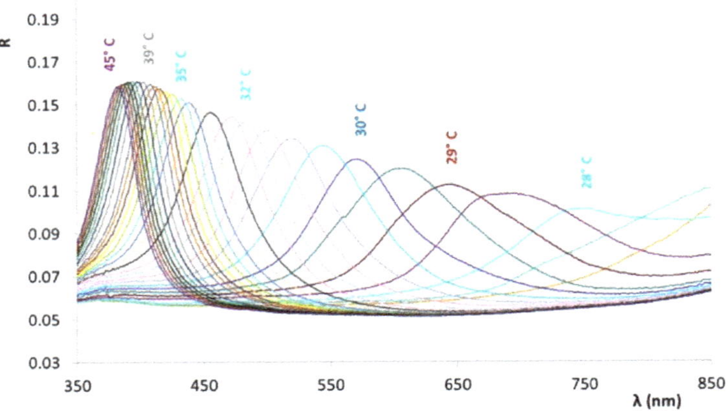

Figure 6. Reflectance spectra of triple-printed TLC layer on coated black paper, measured by the Lambda 950 spectrometer. The temperature of some reflectance spectra is given with the same colour as the corresponding spectrum.

These measurements were evaluated in CIELAB colour space using the (a^*,b^*) plot (Figure 7). The full loop was obtained for all three samples. As expected, the colour play effect is the faintest for the single-printed sample and the strongest for the triple-printed one. This is shown by difference in size of the corresponding colour loops in the (a^*,b^*) plot. The differences between the three samples are the greatest in the green-blue part ($a^* < 0$), where they occur for all the temperatures in the corresponding range. In addition, the triple-printed sample also yields substantial effects in the yellow ($b^* > 0$) parts of the (a^*,b^*) graph (Figure 7).

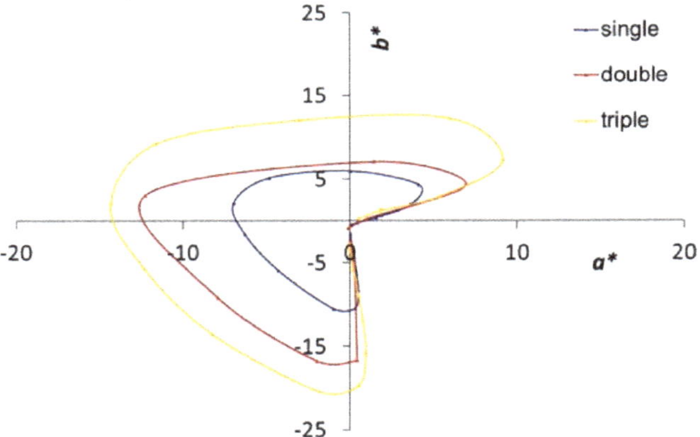

Figure 7. CIELAB colour values of TLC ink, single, double, and triple printed on black coated paper, presented in (a^*,b^*) plot.

The CIELAB colour values include a^* (red-green), b^* (yellow-blue), and L^* (lightness) values. The a^* and b^* values were analysed by (a^*,b^*) plots (Figures 3 and 7), while L^* is examined by the $L^*(T)$ plot (Figure 8). Each curve extends from 27 to 45 °C, where the reflection peak appears in visible, justifying the colorimetric measurements within this range. A single maximum occurs in each $L^*(T)$ curve in the green region, where the colour has the highest lightness L^* (Table 1). The effect is nearly proportional to the thickness of the TLC layer, corresponding to the larger amount of the thermochromic material. At the temperature of 35 °C, the $L^*(T)$ reaches the peak for single- and triple-printed samples, but for the double-printed one, it differs by 0.5 °C, reaching the peak at 31 °C (Table 1). This minute difference could be explained by possible incongruities during measurement.

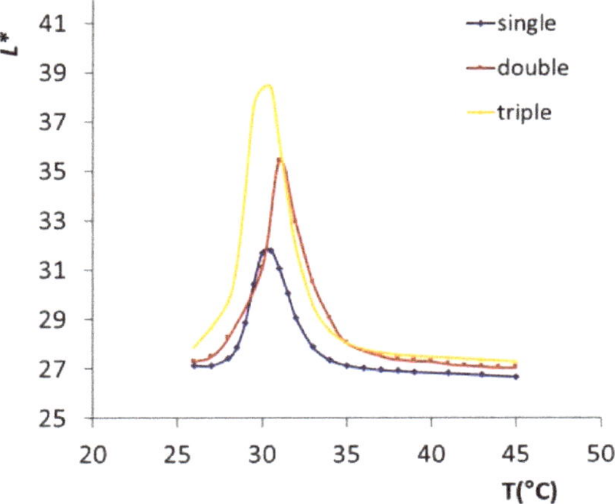

Figure 8. Temperature dependence of CIELAB lightness values L^* measured for single-, double- and triple-printed TLC samples (on black coated paper).

Table 1. Properties of TLC ink printed in single, double and triple layers over the black coated substrate: temperature at which the $L^*(T)$ curve has its maximum is denoted by $T(L^*_{max})$, and its intensity by ΔL^*_{max}. See also Figure 8.

TLC Sample	Thickness (μm)	T (L^*_{max}) (°C)	ΔL^*_{max}
single	27.6	30.5	4.28
double	34.2	31	9.01
triple	46.8	30.5	11.01

Reflectance spectra obtained by both integrating spheres (Figures 1 and 2) were analysed in terms of the position of the reflectance peak λ_{max} as a function of temperature (Figure 9). Here, the $\lambda_{max}(T)$ curves coincide almost completely. The λ_{max} shifts towards shorter wavelengths exponentially with an increase in temperature, which is in accordance with the findings reported in the literature, revealing the change in the pitch length in the TLC material [4,5,7,9].

The position of the reflection peak was analysed in more detail, also taking into account the surface properties of the two substrates and the thickness of the TLC layers. For this purpose, the results were analysed for four samples, namely single, double, and triple layers on black coated substrate and double layer on black uncoated substrate. The $\lambda_{max}(T)$ curves of these samples coincide in full (Figure 10). The only exception is the result at 32.5 °C obtained for the double-printed layer on uncoated black paper, which differs slightly from the other three, but this can be attributed to experimental error.

These results demonstrate that the position of the peak in reflection spectra of TLC samples is unrelated to the spectrometer and integrating sphere used for the measurements (Figure 9). Moreover, the results do not depend either on the substrate and its surface properties (coated or uncoated, i.e., gloss or matt surface) or on the thickness of the TLC layer (Figure 10).

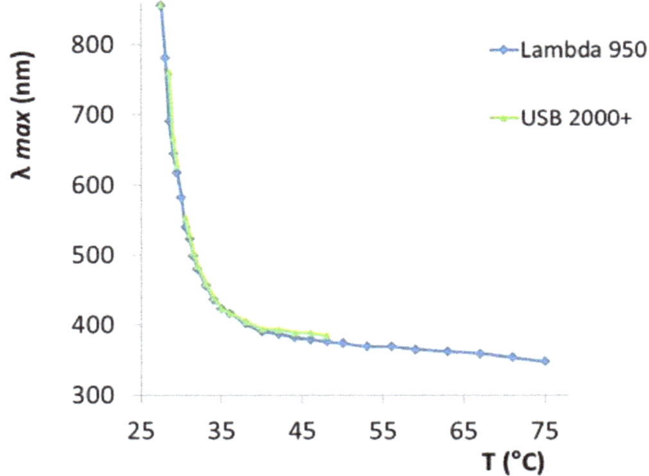

Figure 9. Position of the reflectance peak dependent on temperature as measured by Lambda 950 and USB 2000+ spectrometers (referring to Figures 1 and 2). The corresponding spectra were measured for double layers on black uncoated paper.

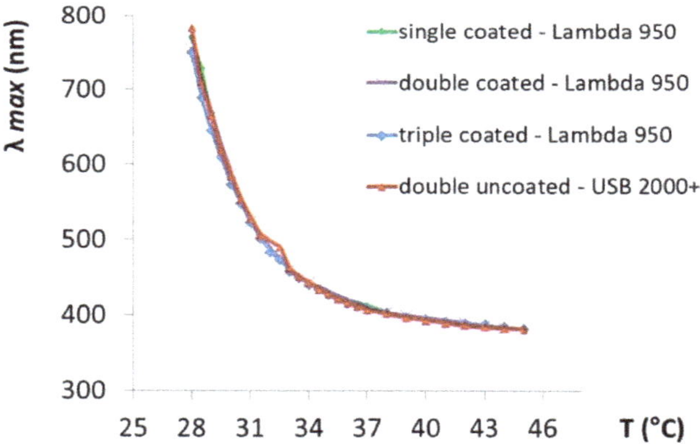

Figure 10. Position of the reflectance peak dependent on the temperature as measured by Lambda 950 and USB 2000+ spectrometers. The reflectance spectra were measured for multiple layers on black coated paper and for double layer on black uncoated paper.

4. Conclusions

Reflectance spectra as a function of temperatures of the TLC ink printed on coated and uncoated black substrate in single-, double- and triple-ink layers were measured by two different spectrometers—the Lambda 950 UV-VISNIR full-size device and the portable fibre-based USB 2000+ Ocean Optics. Both spectrometers use integrating spheres with identical measurement geometry (8°:di), but have different sphere diameter and sample port opening. The reflectance spectra obtained by these devices are highly similar to each other, though not identical. In both cases, the single reflection peak appears as a result of functional material being in the chiral nematic/cholesteric phase. The peak shifts along the visible range due to the increase of temperature and becomes narrower and more intense. When presented as CIELAB colour values, the differences are even more emphasized, despite a very similar shape of the curve obtained in the same trend of curve formation. The differences between the curves for different ink layers are even more significant, with single-layer samples showing the faintest colour play effect visible in the CIELAB diagram, and the triple-layer sample showing the strongest one, covering the largest CIELAB loop. None of these results can prove that the same TLC ink was used throughout the experiment. However, when the wavelengths at which the reflectance peak appears are extracted—the λ_{max}—and presented as variables of the temperature, identical curves are obtained irrespective of the ink layers and substrate properties, and the same is true for both measuring devices. This is an authentic characteristic of TLC ink, independent of any other experimental parameter. We have proposed to call this feature the communication mechanism.

According to our studies, which have been ongoing since 2012, only the TLC inks require the specific procedure described, where only the temperature dependence of the peak in the reflectance spectra is the true identifier of the communication mechanism [7,14].

Author Contributions: Conceptualization, M.K.G., M.S.J. and B.L.; methodology, M.K.G. and M.S.J.; software, M.K.G. and M.S.J.; validation, M.K.G.; formal analysis, M.K.G. and M.S.J.; investigation, M.K.G. and M.S.J.; resources, M.K.G. and M.S.J.; data curation, M.K.G. and M.S.J.; writing—original draft preparation, M.K.G. and M.S.J.; writing—review and editing, B.L.; visualization, M.S.J. and M.K.G.; supervision, M.K.G. All authors have read and agreed to the published version of the manuscript.

Funding: This research was partially funded by the Slovenian Research Agency, grant no. P2-0393.

Acknowledgments: The authors appreciate the assistance of the Sitopapir d.o.o. print shop in Zagreb, Croatia, for printing samples and for technical support. Maja Strižić Jakovljević acknowledges CMEPIUS (The Centre of the Republic of Slovenia for Mobility and European Educational and Training Programmes) and the Slovenian Ministry of Education, Science, Culture, and Sports for a 3-month bilateral scholarship grant. The authors also acknowledge the financial support of COST FP 1104 and University of Zagreb for financial support in scientific research, titled "Multispectral imaging system for thermochromic prints based on liquid crystals".

Conflicts of Interest: The authors declare no conflict of interest.

References

1. Seeboth, A.; Lotzch, D. *Thermochromic and Thermotropic Materials*, 1st ed.; Jenny Stanford Publishing: New York, NY, USA, 2013; ISBN 9788578110796.
2. Seeboth, A.; Lotzch, D. *Thermochromic Phenomena in Polymers*; Smithers Rapra Technology Limited: Shrewsbury, UK, 2008; ISBN 9781847351128.
3. White, M.A.; Leblanc, M. Thermochromism in Commercial Products. *J. Chem. Educ.* **1999**, *76*, 1201–1205. [CrossRef]
4. Hallcrest Handbook of Thermochromic Liquid Crystal Tecnology. Available online: https://www.hallcrest.com/DesktopModules/Bring2mind/DMX/Download.aspx?Command=Core_Download&EntryId=280&language=en-US&PortalId=0&TabId=163 (accessed on 15 February 2021).
5. Christie, R.M.; Bryant, I.D. An evaluation of thermochromic prints based on microencapsulated liquid crystals using variable temperature colour measurement. *Color. Technol.* **2005**, *121*, 187–192. [CrossRef]
6. Hallcrest, L. Color Change Basics, Microencapsulation. Available online: http://www.hallcrest.com/color-change-basics/microencapsulation (accessed on 15 February 2021).
7. Jakovljević, M.; Lozo, B.; Klanjšek Gunde, M. Spectroscopic evaluation of the colour play effect of thermochromic liquid crystal printing inks. *Color. Technol.* **2017**, *133*, 81–87. [CrossRef]
8. Jakovljević, M.; Lozo, B.; Gunde, M.K.; Arts, G.; Arts, G. Packaging added value solutions by Thermochromic Liquid Crystal-based printed labels. In Proceedings of the Printing for Fabrication, IS&T, Manchester, UK, 12–16 September 2016; Volume 2016, pp. 325–327.
9. Shi, J.; Liu, S.; Zhang, L.; Yang, B.; Shu, L.; Yang, Y.; Ren, M.; Wang, Y.; Chen, J.; Chen, W.; et al. Smart Textile-Integrated Microelectronic Systems for Wearable Applications. *Adv. Mater.* **2020**, *32*, 1901958. [CrossRef] [PubMed]
10. Jang, J.; Oh, B.; Jo, S.; Park, S.; An, H.S.; Lee, S.; Cheong, W.H.; Yoo, S.; Park, J.-U. Human-Interactive, Active-Matrix Displays for Visualization of Tactile Pressures. *Adv. Mater. Technol.* **2019**, *4*, 1900082. [CrossRef]
11. Huang, Q.; Zhu, Y. Printing Conductive Nanomaterials for Flexible and Stretchable Electronics: A Review of Materials, Processes, and Applications. *Adv. Mater. Technol.* **2019**, *4*, 1800546. [CrossRef]
12. Sage, I. Thermochromic liquid crystals. *Liq. Cryst.* **2011**, *38*, 1551–1561. [CrossRef]

13. Klanjšek Gunde, M.; Friškovec, M.; Kulčar, R.; Hauptman, N. Functional properties of the leuco dye-based thermochromic printing inks. In Proceedings of the TAGA 63rd Annual Technical Conference, Pittsburgh, PA, USA, 6–9 March 2011.
14. Jakovljević, M.; Kulčar, R.; Tomašegović, D.; Friškovec, M.; Klanjšek, G. Marta Colorimetric description of thermochromic printing inks. *Acta Graph.* **2017**, *28*, 7–14.
15. Strižić Jakovljević, M.; Lozo, B.; Klanjšek Gunde, M. The properties of printing substrates required for thermochromic liquid-crystal printing inks. *J. Print Media Technol. Res.* **2018**, *4*, 165–170. [CrossRef]
16. Jakovljević, M.; Friškovec, M.; Gunde, M.K.; Lozo, B. Optical properties of thermochromic liquid crystal printing inks. In Proceedings of the 5th International Scientific Conference "Printing Future Days 2013", VWB—Verlag für Wissenschaft und Bildung, Berlin, Germany, 10–12 September 2013; pp. 169–173.
17. Kakade, V.U.; Lock, G.D.; Wilson, M.; Owen, J.M.; Mayhesw, J.E. Accurate heat transfer measurements using thermochromic liquid crystal. Part 1: Calibration and characteristics of crystals. *Int. J. Heat Fluid Flow* **2009**, *30*, 939–949. [CrossRef]
18. Abdullah, N.; Abu Talib, A.R.; Jaafar, A.A.; Mohd Salleh, M.A.; Chong, W.T. The basics and issues of Thermochromic Liquid Crystal Calibrations. *Exp. Therm. Fluid Sci.* **2010**, *34*, 1089–1121. [CrossRef]
19. Sefar, A.G. *Handbook for Screen Printers*; Sefar AG Printing Division: Hong Kong, China, 1999.
20. Kulčar, R.; Friškovec, M.; Hauptman, N.; Vesel, A.; Gunde, M.K. Colorimetric properties of reversible thermochromic printing inks. *Dye Pigments* **2010**, *86*, 271–277. [CrossRef]

Article

Beam Shaping for Wireless Optical Charging with Improved Efficiency

Lei Tian, Jiewen Nie and Haining Yang *

Department of Electronic Science and Engineering, Southeast University, 2 Sipailou, Xuanwu District, Nanjing 210018, China; 220201570@seu.edu.cn (L.T.); 220201566@seu.edu.cn (J.N.)
* Correspondence: h.yang@seu.edu.cn

Abstract: Optical wireless charging is a nonradiative long-distance power transfer method. It may potentially play an important role in certain scenarios where access is challenging, and the radio frequency power transfer is less efficient. The divergence of the optical beam over distances is a key limiting factor for the efficiency of any wireless optical charging system. In this work, we propose and experimentally demonstrate a holographic optical beam shaping system that can restrict the divergence of the optical beam. Our experimental results showed up to 354.88% improvement in the charging efficiency over a 10 m distance.

Keywords: optical wireless power transfer; beam shaping; phase modulation

Citation: Tian, L.; Nie, J.; Yang, H. Beam Shaping for Wireless Optical Charging with Improved Efficiency. *Crystals* **2021**, *11*, 970. https://doi.org/10.3390/cryst11080970

Academic Editors: Akihiko Mochizuki, Kohki Takatoh and Jun Xu

Received: 15 July 2021
Accepted: 15 August 2021
Published: 17 August 2021

1. Introduction

Wireless charging [1–4] has attracted considerable attention in recent years due to its operational flexibility; this is especially true for underwater environments as the design of the charging cable and sockets can be complicated and often impractical. Wireless charging can be implemented in various ways. Magnetic coupling charging [5,6] can transmit energy at the kilowatt level with high efficiency. However, it only supports wireless charging within 1 m as the power declines rapidly with the increasing charging distance [7,8]. The transmission distance of microwave charging is much longer [9]. However, its charging efficiency is limited due to the quick divergence of the microwave beams. The propagation loss of microwaves is also extremely high in the water. Therefore, it is not suitable for underwater applications [3,10]. Wireless optical charging has gained considerable interest in recent years due to its flexibility, strong directivity, and support for long transmission ranges [11]. It is also free of electromagnetic interference. In addition, blue and green wavelengths correspond to the low attenuation window of the seawater, which make them suitable for underwater applications [12]. The sources for wireless optical charging are laser diode [13,14], light-emitting diodes (LED) [15,16], vertical-cavity surface-emitting laser (VCSEL) [17], etc. The performance of the wireless optical charging system has been optimised in previous works. Most of these works mainly focused on fully exploiting the power processing capacity and conversion efficiency of laser and PV cells. The operating point of the laser can be tuned to improve the electrical-to-optical efficiency [18,19]. The material and temperature of the photovoltaic (PV) cell can also be optimised for higher efficiency [20]. However, the divergence of the optical beam over distances [21] may be a key factor for the system's overall efficiency. To our knowledge, this issue has not been properly addressed so far.

In this work, we designed and demonstrated a holographic optical beam shaping system that could improve the overall charging efficiency. In our demonstration, the divergence of the optical beam was restricted for different charging distances. The power on the PV cell significantly increased in both free space and water compared with the conventional system. Our results showed that this technique has great potential for the application of underwater wireless charging.

2. System Setup

Figure 1 shows the design of our proposed optical launching unit with the beam shaping capability. A pigtailed single-mode laser diode (Thorlabs LP450-SF15, Newton, NJ, USA) operating at 450 nm was used as the light source. Then a fibre-coupled collimator (Thorlabs F671FC-405, Newton, NJ, USA) was used to slow the divergence of the laser beam. The wavefront of the collimated beam was spatially modulated by a phase-only liquid crystal on silicon (LCOS) device (CamOptics COVIS-2K, Cambridge, UK). This LCOS device has a bit depth of 8 bit, i.e., 256 unique phase levels between 0 and 2.5π. The phase flicker of the LCOS was optimised [22,23]. The reflectance of the LCOS device was ~80%; however, it can be improved by applying a dielectric mirror coating on the LCOS backplane [24]. The desired beam can be generated with the Fourier transform lens (L1, f1 = 500 mm) in the system. A 4f magnification system based on L2 (f2 = 250 mm) and L3 (f3 = 500 mm) were used to further enlarge the shaped beam. For the proof-of-concept demonstration, a beam splitter (BS) was used in the current system; this can be replaced with a prism to further increase the energy efficiency of this optical launching unit [25].

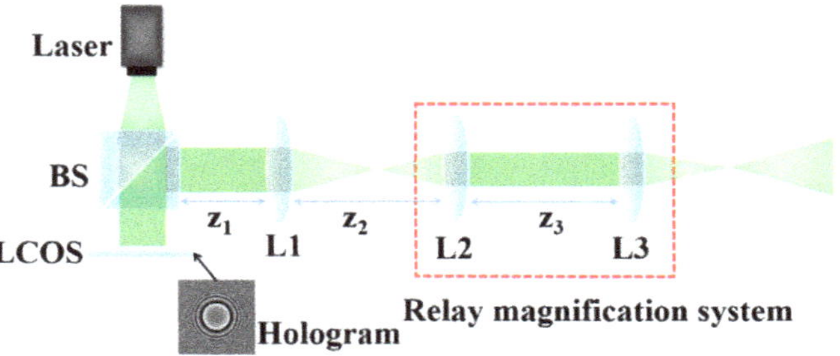

Figure 1. Setup of the optical launching unit.

Two types of beam shaping techniques were investigated in this work. The first one uses holograms that exhibit radially symmetric cubic phase profiles, which is described by the following equation [26,27]:

$$P = angle\left(e^{-ib|r|^3}\right) \tag{1}$$

where r is the radial distance and b is a coefficient optimised for different charging distances.

The second type utilises holograms that follow the Fresnel lens shape [28,29], which is described by:

$$P = angle\left(e^{-ibr^2}\right) \tag{2}$$

where r is the radial distance and b is related to focal length of the Fresnel lens. Again, the value of b can be optimised for different charging distances.

Figure 2 shows the general architecture of our testbed. A pigtailed single-mode laser diode (Thorlabs LP450-SF15, Newton, NJ, USA) operating at 450 nm was driven by Thorlabs LDM9LP to provide a constant DC of 35 mA with an output power of approximately 4.37 mW. This driver unit also integrated a thermoelectric cooler module to maintain the laser performance during the experiment. Then the laser was fed from the single-mode fibre into an optical launching unit for collimation and beam shaping before entering the channel. At the receiver side, a low-cost PV cell was used to collect the incident optical power. In this way, the received optical power was converted into electric power to charge a target device. A source meter (KEITHLEY-2400, Portland, OR, United States) was used to collect the voltage and current passing through the cell. In this work, we tested wireless optical charging in both free spaces and water environments.

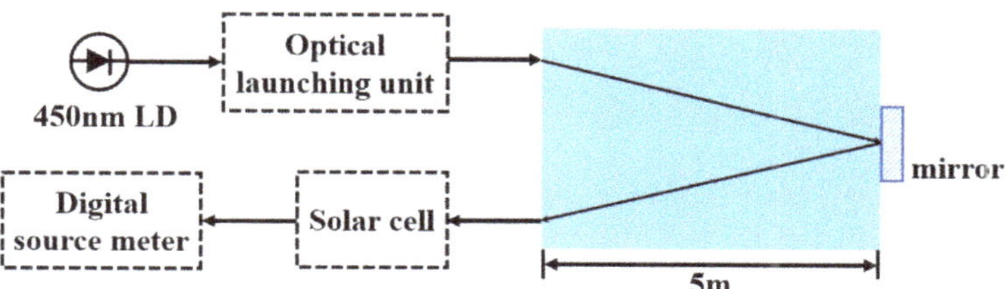

Figure 2. General architecture of the testbed.

3. Results

First, simulations based on the Fourier optics theory [30] were conducted to investigate the performance of the optical launching unit. It should be noted that a 4f system of L2 and L3 had a magnification factor of 2. The transmission medium was assumed to be homogeneous and had a refractive index of 1. The results were shown in Figure 3. Figure 3a corresponded to the beam propagation behaviour when the optical beam shaping was not applied. It diverged with a fixed angle while propagating through the medium. Figure 3b corresponded to the case when a radially symmetric cubic phase modulation applied with $b = 4.95 \times 10^{11}$. The beam showed a strong self-focusing characteristic within the first ~20 m before diverging. The highest intensity of the modulated beam was 7.8 dB higher than the Gaussian beam at the same distance. This simulation results when the Fresnel lens phase hologram with a b value of 1.25×10^8 was illustrated in Figure 3c. In this case, the position of the maximum light intensity was similar to Figure 3b. A further improvement of 4.01 dB was achieved. After passing through the focal point, the light field diverged as a Gaussian beam. The simulation results showed that the Fresnel phase hologram was more effective than the radially symmetric cubic phase hologram.

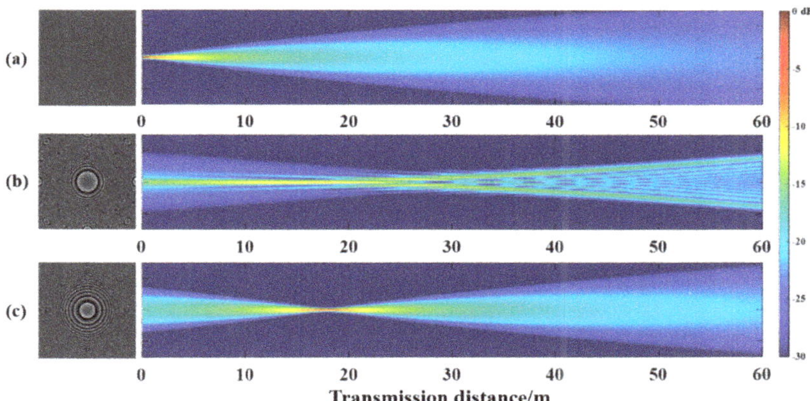

Figure 3. The phase holograms and simulated characteristics of the launched beams (**a**) without beam shaping; (**b**) with the cubic modulation beam shaping when $b = 4.95 \times 10^{11}$; (**c**) with the Fresnel modulation beam shaping when $b = 1.25 \times 10^8$.

Subsequently, the impact of b values on the system of beam propagation characteristics was analysed. As shown in Figure 4, the focusing position of the beam increased with the value of b when the cubic modulation was implemented. Although the peak light intensity decreased for the longer transmission distances, it was always significantly higher than the standard Gaussian beam without beam shaping.

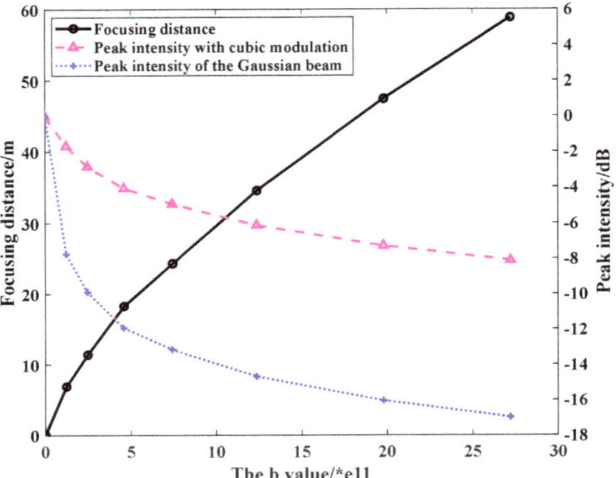

Figure 4. The trend of focusing distance and intensity with b value when loading the cubic modulation hologram.

Figure 5 showed corresponding results when the Fresnel modulation was used. The trend was similar to that observed in Figure 4. The notable difference was that the intensity attenuation was very small at different focusing distances with Fresnel modulation. Compared with cubic modulation, Fresnel modulation can increase the light intensity more significantly.

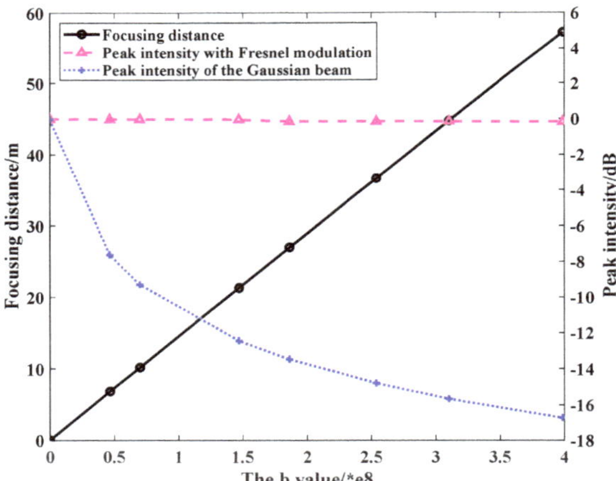

Figure 5. The trend of focusing distance and intensity with b value when loading the Fresnel modulation hologram.

In our experiment, the voltage-current curve of the PV cell was measured by a source metre. The measurement was conducted when the PV cell was placed 10 m away from our launching unit and the beam shaping was not applied. The results were plotted in Figure 6. The maximum power generated in this specific setting was 1.91 μW with a load resistance of 5250 Ω.

Subsequently, we measured the maximum electrical power generated by the PV cell under different beam shaping configurations. The experiment was first conducted in the free space between 4 m and 10 m. The value of *b* was optimised for each transmission distance. Figure 7 illustrated the maximum power generated at each plane. It can be seen that the generated power decreased over the distance. However, the drop was most significant when no beam shaping was applied. When the Fresnel phase modulation was used, the decrease in power was the slowest. The value of b used in each measurement was listed in Table 1.

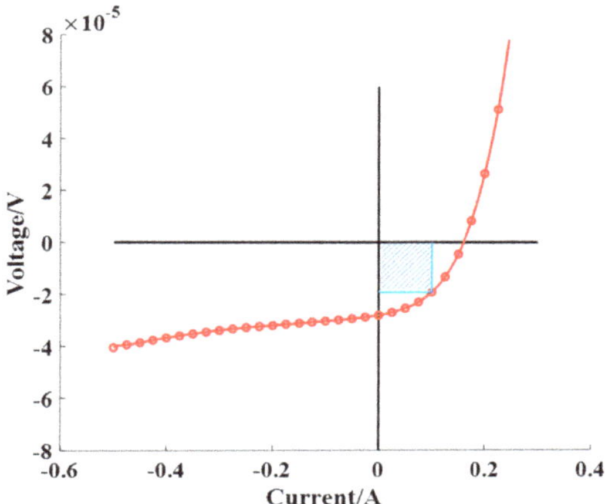

Figure 6. Voltage vs. current curve of the PV cell.

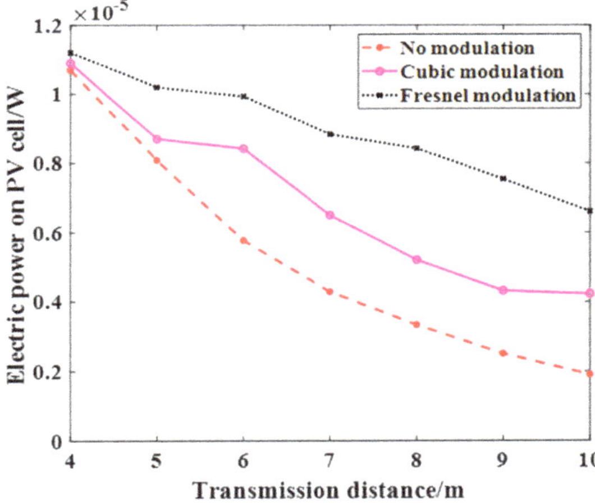

Figure 7. The maximum electrical power obtained by source metre conversion at different planes.

Table 1. The values of b used in the experiments.

Focusing Distance/m	Cubic Modulation/($\times 10^{11}$)	Fresnel Modulation/($\times 10^{8}$)
4	0.7324	0.4189
5	1.1719	0.6981
6	1.7578	0.9774
7	2.4902	1.2566
8	3.8086	1.5359
9	5.2734	1.8151
10	8.2031	2.0944

Figure 8 showed the relative improvement in the charging efficiency at each plane when compared with the results measured without beam shaping. At 10 m, the charging efficiency increased by 246% when the Fresnel modulation was used.

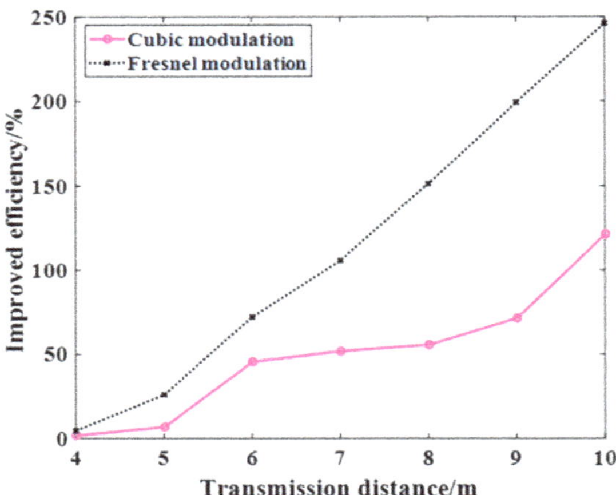

Figure 8. The efficiency of power improvement.

We then conducted the same experiment in a water tank. The attenuation coefficient of water in the tank was measured as ~0.25/m, which was similar to the Jerlov II water conditions. The experimental results underwater were nearly the same as those in free space. Since the tank was 5 m long, only the situations at 5 m and 10 m were tested. At 5 m, the radially symmetric cubic phase modulation increased power by 21.33% over the Gaussian and the Fresnel phase modulation by 26.22%. At 10 m, the former increased by 176.24% and the latter by 354.88%. At a longer distance, the beam size without shaping became significantly larger. The beam shaping techniques were able to concentrate the optical power at an arbitrary receiver plane. Therefore, its effect became more apparent at a longer distance, improving the overall system efficiency.

4. Conclusions

In this work, we demonstrated two holographic optical beam shaping techniques that could improve the optical energy transfer efficiency in free spaces and underwater environments. In a proof-of-concept experiment, both techniques demonstrated focusing capability at different distances. The beam shaping based on the Fresnel lens modulation performed better than the radially symmetric cubic phase modulation. We believe the holographic optical beam shaping technique has the potential for application in underwater wireless optical charging systems, where the accessibility is poor, and the RF transmission is inefficient. Although the power currently generated in this work is low, it could be significantly improved by using more powerful laser sources and more efficient PV cells. Using the beam splitter in our current configuration also limited the overall efficiency of the system. However, it can be easily solved by replacing the beam splitter with a prism, as previously mentioned. Alternatively, transmissive phase-only spatial light modulators can also be used.

Author Contributions: Conceptualization, H.Y.; methodology, H.Y.; validation, L.T. and J.N.; formal analysis, L.T., J.N. and H.Y.; writing—original draft preparation, L.T.; writing—review and editing, L.T., J.N. and H.Y.; supervision, H.Y.; funding acquisition, H.Y. All authors have read and agreed to the published version of the manuscript.

Funding: This research and APC were funded by the Natural Science Foundation of Jiangsu Province (BK20200351); Jiangsu Special Professorship.

Acknowledgments: The authors would like to thank Chaogang Lou for providing the source metre.

Conflicts of Interest: The authors declare no conflict of interest.

References

1. Garnica, J.; Chinga, R.A.; Lin, J. Wireless Power Transmission: From Far Field to Near Field. *Proc. IEEE* **2013**, *101*, 1321–1331. [CrossRef]
2. Fisher, T.M.; Farley, K.B.; Gao, Y.; Bai, H.; Tse, Z.T.H. Electric vehicle wireless charging technology: A state-of-the-art review of magnetic coupling systems. *Wirel. Power Transf.* **2014**, *1*, 87–96. [CrossRef]
3. Lu, X.; Wang, P.; Niyato, D.; Kim, D.I.; Han, Z. Wireless Charging Technologies: Fundamentals, Standards, and Network Applications. *IEEE Commun. Surv. Tutor.* **2016**, *18*, 1413–1452. [CrossRef]
4. Zhou, J.; Zhang, B.; Xiao, W.; Qiu, D.; Chen, Y. Nonlinear Parity-Time-Symmetric Model for Constant Efficiency Wireless Power Transfer: Application to a Drone-in-Flight Wireless Charging Platform. *IEEE Trans. Ind. Electron.* **2019**, *66*, 4097–4107. [CrossRef]
5. Dai, J.; Ludois, D.C. Single active switch power electronics for kilowatt scale capacitive power transfer. *IEEE J. Emerg. Sel. Top. Power Electron.* **2014**, *3*, 315–323.
6. Sakai, N.; Itokazu, D.; Suzuki, Y.; Sakihara, S.; Ohira, T. One-kilowatt capacitive Power Transfer via wheels of a compact Electric Vehicle. In Proceedings of the 2016 IEEE Wireless Power Transfer Conference (WPTC), Aveiro, Portugal, 5–6 May 2016; pp. 1–3.
7. Jawad, A.M.; Nordin, R.; Gharghan, S.K.; Jawad, H.M.; Ismail, M. Opportunities and Challenges for Near-Field Wireless Power Transfer: A Review. *Energies* **2017**, *10*, 1022. [CrossRef]
8. Abou Houran, M.; Yang, X.; Chen, W. Magnetically Coupled Resonance WPT: Review of Compensation Topologies, Resonator Structures with Misalignment, and EMI Diagnostics. *Electronics* **2018**, *7*, 296. [CrossRef]
9. Sasaki, S.; Tanaka, K.; Maki, K. Microwave Power Transmission Technologies for Solar Power Satellites. *Proc. IEEE* **2013**, *101*, 1438–1447. [CrossRef]
10. Shimamura, K.; Sawahara, H.; Oda, A.; Minakawa, S.; Mizojiri, S.; Suganuma, S.; Mori, K.; Komurasaki, K. Feasibility study of microwave wireless powered flight for micro air vehicles. *Wirel. Power Transf.* **2017**, *4*, 146–159. [CrossRef]
11. Jin, K.; Zhou, W. Wireless Laser Power Transmission: A Review of Recent Progress. *IEEE Trans. Power Electron.* **2019**. *34*, 3842–3859. [CrossRef]
12. Khalighi, M.; Gabriel, C.; Hamza, T.; Bourennane, S.; Léon, P.; Rigaud, V. Underwater wireless optical communication; recent advances and remaining challenges. In Proceedings of the 2014 16th International Conference on Transparent Optical Networks (ICTON), Graz, Austria, 6–10 July 2014; pp. 1–4.
13. Kim, S.-M.; Kwon, D. Transfer Efficiency of Underwater Optical Wireless Power Transmission Depending on the Operating Wavelength. *Curr. Opt. Photonics* **2020**, *4*, 571–575.
14. Duncan, K.J. Laser based power transmission: Component selection and laser hazard analysis. In Proceedings of the 2016 IEEE PELS Workshop on Emerging Technologies: Wireless Power Transfer (WoW), Knoxville, TN, USA, 4–6 October 2016; pp. 100–103.
15. Zhou, Y.; Miyamoto, T. 200 mW-class LED-based optical wireless power transmission for compact IoT. *Jpn. J. Appl. Phys.* **2019**, *58*, SJJC04. [CrossRef]
16. Uchiyama, N.; Yamada, H. Proposal and demonstration of LED optical wireless power-transmission systems for battery-operated small electronic devices. *Jpn. J. Appl. Phys.* **2020**, *59*, 124501. [CrossRef]
17. Hirota, M.; Iio, S.; Ohta, Y.; Niwa, Y.; Miyamoto, T. Wireless power transmission between a NIR VCSEL array and silicon solar cells. In Proceedings of the 2015 20th Microoptics Conference (MOC), Fukuoka, Japan, 25–28 October 2015; pp. 1–2.
18. Kim, S.-M.; Choi, J.; Jung, H. Experimental demonstration of underwater optical wireless power transfer using a laser diode. *Chin. Opt. Lett.* **2018**, *16*, 80101. [CrossRef]
19. Höhn, O.; Walker, A.W.; Bett, A.W.; Helmers, H. Optimal laser wavelength for efficient laser power converter operation over temperature. *Appl. Phys. Lett.* **2016**, *108*, 241104. [CrossRef]
20. Valdivia, C.E.; Wilkins, M.M.; Bouzazi, B.; Jaouad, A.; Aimez, V.; Arès, R.; Masson, D.P.; Fafard, S.; Hinzer, K. Five-volt vertically-stacked, single-cell GaAs photonic power converter. In Proceedings of the SPIE, San Francisco, CA, USA, 7–12 February 2015; Volume 9358.
21. Raible, D.E. *High Intensity Laser Power Beaming for Wireless Power Transmission*; Cleveland State University: Cleveland, OH, USA, 2008.
22. Yang, H.; Chu, D.P. Phase flicker optimisation in digital liquid crystal on silicon devices. *Opt. Express* **2019**, *27*, 24556–24567. [CrossRef] [PubMed]
23. Yang, H.; Chu, D.P. Phase flicker in liquid crystal on silicon devices. *J. Phys. Photonics* **2020**, *2*, 032001. [CrossRef]
24. Yang, H.; Chu, D.P. Digital phase-only liquid crystal on a silicon device with enhanced optical efficiency. *OSA Contin.* **2019**, *2*, 2445–2459. [CrossRef]
25. Liu, S.; Qi, S.; Zhang, Y.; Li, P.; Wu, D.; Han, L.; Zhao, J. Highly efficient generation of arbitrary vector beams with tunable polarization, phase, and amplitude. *Photonics Res.* **2018**, *6*, 228–233. [CrossRef]
26. Zhang, Z.; Liang, X.; Goutsoulas, M.; Li, D.; Yang, X.; Yin, S.; Xu, J.; Christodoulides, D.N.; Efremidis, N.K.; Chen, Z. Robust propagation of pin-like optical beam through atmospheric turbulence. *APL Photonics* **2019**, *4*, 76103. [CrossRef]

27. Nie, J.; Tian, L.; Yue, S.; Zhang, Z.; Yang, H. Advanced Beam Shaping for Enhanced Underwater Wireless Optical Communication. In Proceedings of the 2021 Optical Fiber Communications Conference and Exhibition (OFC), San Francisco, CA, USA, 6–10 June 2021.
28. Nie, J.; Tian, L.; Wang, H.; Chen, L.; Li, Z.; Yue, S.; Zhang, Z.; Yang, H. Adaptive beam shaping for enhanced underwater wireless optical communication. *Opt. Express* **2021**, *29*, 26404–26417. [CrossRef]
29. Robertson, B.; Zhang, Z.; Yang, H.; Redmond, M.M.; Collings, N.; Liu, J.; Lin, R.; Jeziorska-Chapman, A.M.; Moore, J.R.; Crossland, W.A.; et al. Application of the Fractional Fourier Transform to the Design of LCOS Based Optical Interconnects and Fiber Switches. *Appl. Opt.* **2012**, *51*, 2212–2222. [CrossRef]
30. Goodman, J.W. *Introduction to Fourier Optics*, 2nd ed.; McGraw-Hill Series in Electrical and Computer Engineering; McGraw-Hill: New York, NY. USA, 1996; ISBN 0070242542.

Article

Polymer-Stabilized Blue Phase and Its Application to a 1.5 μm Band Wavelength Selective Filter

Seiji Fukushima [1,*], Kakeru Tokunaga [1], Takuya Morishita [2], Hiroki Higuchi [2], Yasushi Okumura [2], Hirotsugu Kikuchi [2] and Hidehisa Tazawa [3]

1 Graduate School of Science and Engineering, Kagoshima University, Korimoto 1, Kagoshima-shi, Kagoshima 890-0065, Japan; k7054919@kadai.jp
2 Institute for Materials Chemistry and Engineering, Kyushu University, 6-1 Kasuga-koen Kasuga-shi, Fukuoka 816-8580, Japan; morishita.takuya.158@s.kyushu-u.ac.jp (T.M.); higuchi@cm.kyushu-u.ac.jp (H.H.); okumura@cm.kyushu-u.ac.jp (Y.O.); kikuchi@cm.kyushu-u.ac.jp (H.K.)
3 Sumitomo Electric Industries Ltd., 1 Tayacho, Sakae-ku, Yokohama-shi 244-8588, Japan; tazawa-hidehisa@sei.co.jp
* Correspondence: fukushima@eee.kagoshima-u.ac.jp; Tel./Fax: +81-99-285-8438

Abstract: The use of polymer-stabilized blue phase (PSBP) including a tolane-type liquid crystal was investigated to develop a voltage-controlled wavelength selective filter for wavelength-division-multiplexing optical fiber network. It was found that the tolane-type liquid crystal introduction can increase both a blue-phase temperature range and a Kerr coefficient. A Fabry–Perot etalon filled with PSBP functioned as a wavelength selective filter, as expected. The tuning wavelength range was 62 nm although peak transmission was not as high as expected. Numerical analysis suggested that light absorption in transparent electrodes may cause the issue. Minor change to the etalon structure will result in improved performance.

Keywords: liquid crystal; variable wavelength filter; blue phase; stabilized polymer

Citation: Fukushima, S.; Tokunaga, K.; Morishita, T.; Higuchi, H.; Okumura, Y.; Kikuchi, H.; Tazawa, H. Polymer-Stabilized Blue Phase and Its Application to a 1.5 μm Band Wavelength Selective Filter. *Crystals* **2021**, *11*, 1017. https://doi.org/10.3390/cryst11091017

Academic Editors: Kohki Takatoh, Jun Xu and Akihiko Mochizuki

Received: 20 July 2021
Accepted: 23 August 2021
Published: 25 August 2021

Publisher's Note: MDPI stays neutral with regard to jurisdictional claims in published maps and institutional affiliations.

1. Introduction

A wavelength selective filter (WSF) has become an essential device in a wavelength-division-multiplexed (WDM) optical fiber communication system, where multiple color signals are transmitted in parallel in the optical fiber to increase communication throughput [1–4]. Current systems employ wavelengths between 1.3 and 1.6 μm as optical fibers show the smallest absorption and the smallest chromatic dispersion near these wavelengths. The WSFs are necessary both in transmitters and receivers in such systems. WDM systems could be developed by using constant wavelength filters while the systems employing the WSFs have more advantages in design flexibility, low-cost construction and maintenance, and functionality. Furthermore, system researchers and designers impose other requirements such as low-voltage operation, low-power consumption, fast response, and wide temperature range. A large number of WSFs have been developed and some are commercially available [5,6]. It should be noted that the use of liquid crystal (LC) is advantageous in large change in refractive index and low-voltage and low-power operation [7]. Response times of nematic LC are typically in ms order although some LC materials can respond in μs order [8].

Polymer stabilization of LC could extend its application field to optical fiber communication devices. We have succeeded in development of a variable optical attenuator for 1.5 μm optical fiber networks [9]. This paper focuses on the use of polymer-stabilized blue phase (PSBP) for a WSF. We were drawn to conventional—unstabilized—blue phase devices as a viable option due to its non-birefringence at zero electric field, high speed response, and large Kerr effect [10]. However, a major disadvantage of the conventional blue phases is narrow temperature range of the phase. Blue phase is eroded by a chiral nematic phase and an isotropic phase. As a result, blue phase appears only in a few °C at temperatures higher than room temperature. To overcome this problem, we set out to fix blue phase liquid crystal in polymer network. Thus, we achieved a blue-phase temperature range of 47 °C by using tolane-type liquid crystal as host mixture [11]. Another issue that needed to be resolved is to decrease the applied voltage. Typically, a higher voltage operation is necessary due to the short coherent length in orientational order and highly twisted molecular alignment in the conventional devices.

Then, we filled a Fabry–Perot etalon with the PSBP to develop a lower-voltage WSF and succeeded in its demonstration. A tuning wavelength range of 62 nm was experimentally obtained at 120 V or lower voltages. In other words, the required voltage for 7 nm tunability is 13.5 V, assuming an application to access network called NG-PON2 [4].

A molecule structure, a sample preparation, and a phase diagram of the PSBP are described in Section 2. The structure and experimental results are described in Section 3 that includes transmission spectrum of the PSBP-filled WSF. We discuss the characteristics of the developed WSF and suggestions for future improvement in Section 4. We conclude this paper in Section 5.

2. Preparation and Measurement of Mixture

Sample fabrication processes are divided into two: liquid crystal mixture and Fabry–Perot etalon manufacture described below and in Section 3.

First, we describe the principle and mixture method of PSBP. Blue phase appeared in only a few °C between a chiral nematic (N*) phase and an isotropic (Iso) phase in a previous report [10]. It was expected that the temperature range can be expanded by the fixation of LC molecules in polymer-stabilized network. Figure 1a shows a schematic diagram of BP I with double twisted cylinders (blue) and disclination lines (violet) and Figure 1b shows the LC molecules (brown) aligned along the disclination lines. To achieve larger birefringence or low-voltage operation, a tolane-type LC is regarded as one of the strongest candidates as host nematic LC (NLC). Schematics of general tolane-type molecule and TE-2FF are shown in Figure 2a,b, respectively. The use of tolane-type host LC can enlarge the Kerr effect and decrease the operation voltage. We chose two tolane-type LCs: TE-2FF and E8 (Merck, Darmstadt, Germany). Precursors to PSBP consist of the described host LCs, a chiral dopant, a mono-functional monomer, a bi-functional monomer, and a photo initiator. Details of the materials are summarized in Table 1. Then, the precursor mixture was irradiated with 365 nm, 1.5 mW/cm^2 ultra-violet (UV) light to the blue phase for 20 min. Phase transition of the mixture was observed during preparation by using a polarization microscope at a wavelength of 633 nm before and after UV light irradiation. Materials were observed between two glass substrates during fabrication. The injection was done at 110 °C for the TE-2FF:E8 ratio of 1:1 and 1:2 and at 95 °C for the ratio of 1:2.5, 1:3, and 1:4.

We obtained phase diagrams with a parameter of the TE-2FF:E8 ratio, as shown in Figure 3a,b, that were observed before and after UV irradiation, respectively. We describe the characteristics for the cell before UV light irradiation first. The developed materials show BP I phase between N* phase and isotropic phase for all the host mixture ratios. Between the N*-phase and the isotropic phase, a BP I phase appears only in 4.4–4.7 °C, as shown in Figure 3a. The BP I temperature range was independent of the mixture ratio. In contrast, we observed wider temperature range of BP I phase in the sample after UV irradiation or the polymer-stabilized sample, as shown in Figure 3b. The BP I temperature ranges are 10 °C for a ratio of 1:1 and 47 °C for the ratio of 1:4. The temperature range becomes widened as the E8 ratio increases. Furthermore, we observed that the BP I phase appears at a temperature as low as 20 °C, which is regarded as room temperature. The BP I phase appears between the N* phase and the isotropic phase for the cells with TE-2FF:E8 of 1:1 while we find that the BP I phase appears between the crystal phase and isotropic phase for the cells with TE-2FF:E8 ratio of 1:2 to 1:4. The latter cells show no N* phase.

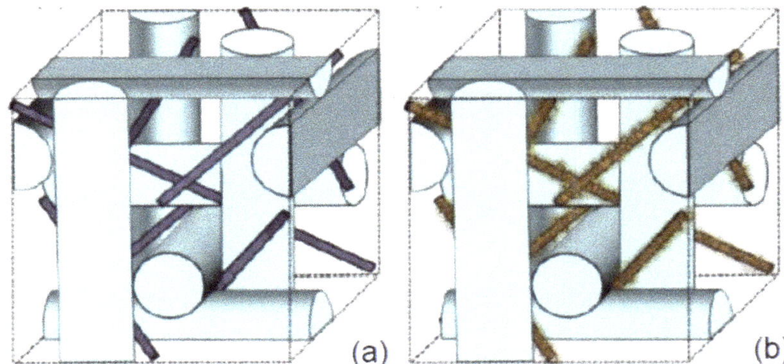

Figure 1. (**a**) Sketch of blue-phase liquid crystal with some disclination lines (blue) and (**b**) polymer-stabilized blue-phase liquid crystal. Polymer network (brown) is constructed along the disclination lines.

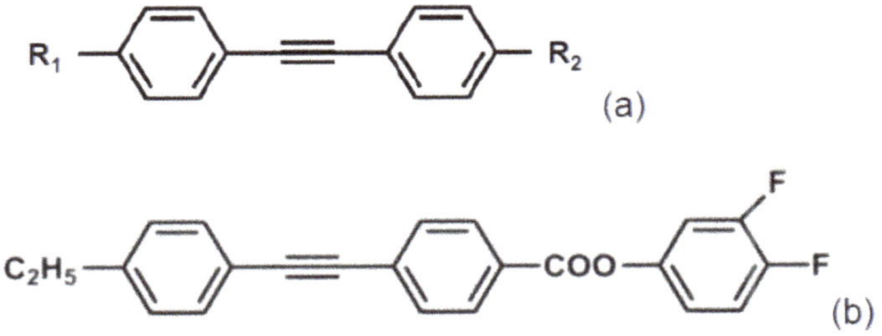

Figure 2. (**a**) Molecule sketch of a general tolane-type liquid crystal and (**b**) TE-2FF.

Table 1. Composition in our blue phases.

Mixture	Chemical Material	wt%
Host LC (92 wt%)	NLC: E8 (Merck)	x
	NLC: TE-2FF	92.5 − x
	Chiral dopant: s-[4′-(hexyloxy)-phenyl-4-carbonyl]-1,4;3,6-dianhydride-D-sorbitol(ISO-(6OBA)2	7.5
Polymer (8 wt%)	Mono-functional monomer: Dodecyl acrylate (C12A, Wako)	38.4
	Bi-functional monomer: 2-methyl-1,4-phenylene-bis(4-(3-(acryloyloxy)propyloxy)benzoate) (RM257, Merck)	57.6
	Photo initiator: 2,2-dimethoxy-2-phenylacetophenone (DMPAP, Aldrich)	4.0

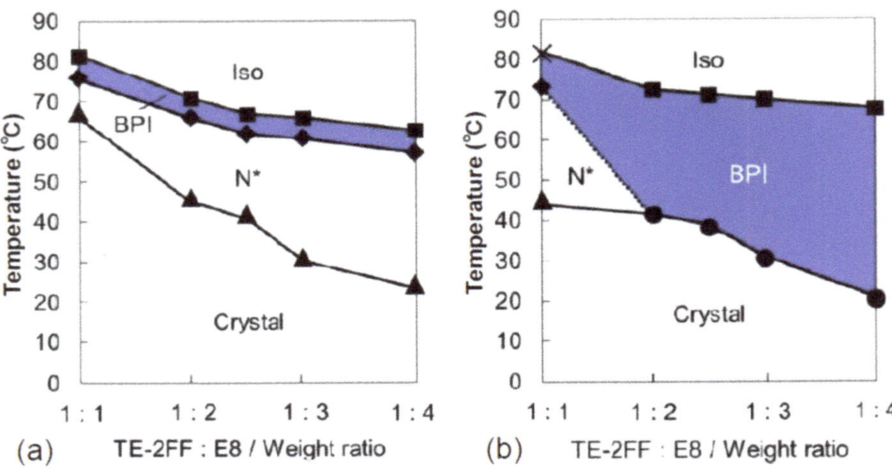

Figure 3. (**a**) Phase diagram of our mixture before UV irradiation and (**b**) after UV irradiation; that is, the mixture is already polymer-stabilized.

We observed the Kerr effect of our new material and conventional, non-tolane 5CB:JC1041XX material, as shown in Figure 4. JC1041XX was available from JNC, Tokyo, Japan. Light transmission was observed with the crossed-Nicol polarized microscope. All the curves seem proportional to the square of sinusoidal function except at the voltage of lower than 40 V, in which case, they reach their peaks, and then decrease. We should note that the voltage at the peak transmission is 216 V for the conventional mixture while we find 97 and 108 V for our new mixture with TE-2FF:E8 ratios of 1:2.5 and 1:3. These new data shows that reduction of the applied voltage is as high as 55 and 50%.

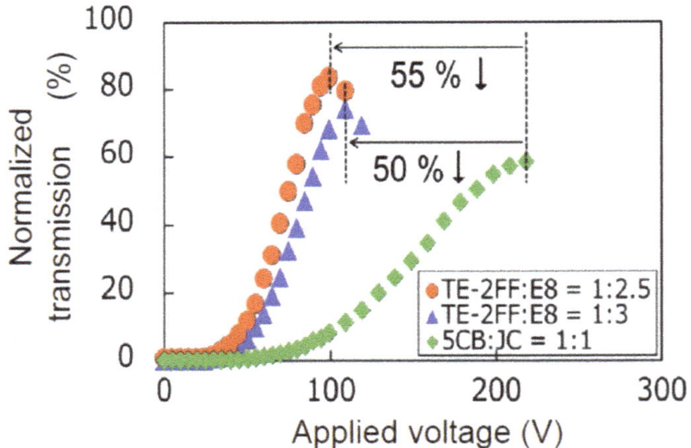

Figure 4. Normalized transmission versus the applied voltage for the tolane-type mixture and the conventional mixture (5CB:JC1041XX). The latter is simply described as 5CB:JC in the figure.

3. Experiments of a Variable Wavelength Filter

A cross-sectional schematic of the WSF is shown in Figure 5. A test cell of WSF was developed as follows: On two glass substrates, we deposited a dielectric mirror of six-pair layers of silicon oxide and titan oxide as high and low refractive index material, respectively, and a layer of indium tin oxide as transparent electrode that was placed inside the mirrors. Each layer of silicon oxide and titan oxide has a thickness (optical length) of quarter of the wavelength of 1.5 μm. We designed the cells, assuming that refractive index of titan oxide is 2.2 and that of silicon oxide is 1.4. The two glass substrates were fixed with 10 μm thick spacers and adhesive, which are not shown in Figure 5 to omit complexity. The precursor mixture was injected into a Fabry–Perot etalon at 110 °C for the TE-2FF:E8 ratio of 1:1 and 1:2 and at 95 °C for the ratio of 1:2.5, 1:3, and 1:4. Then, the cell was irradiated with 365 nm and 1.5 mW/cm^2 UV light for 20 min to be polymer-stabilized in the etalon. Finally, we sealed the cell and bonded electric wires for measurements. The PSBP thickness was estimated to be larger than the spacer thickness of 10 μm and its refractive index was approximately 1.4.

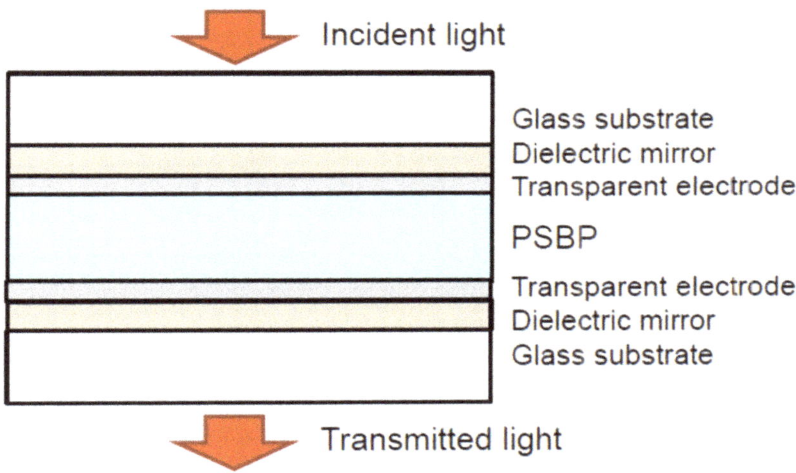

Figure 5. Cross-sectional schematic of the proposed variable wavelength filter.

To test the cells, we employed an optical spectrum analyzer with a built-in broadband light source. Light from the light source was fed through a single-mode optical fiber to be a parallel beam after passing through a focusing lens. The light beam was incident on the cell and concentrated and fed into the optical spectrum analyzer with another lens and another single-mode optical fiber. The measurement configuration was setup in a box so that the optical measurement was measured stably. No matching oil was employed in the setup. Figure 6 shows measured transmission spectra of the developed device at the applied voltages of 0 (blue), 40 (green), 80 (yellow), and 120 (red) V. The applied electric signal was alternative current at an audio frequency. The light was not polarized and the temperature was 27 °C. We evaluated only the cell with the TE-2FF:E8 of 1:4. At 0 V, the spectrum shows a curve similar to the square of sinusoidal function. The peak transmission wavelength decreases clearly as the applied voltage increases. The developed device functioned successfully as a WSF; however, the maximum transmission is as low as 2.2% and the half-bandwidth is 23 nm. The low transmission suggests that absorption may occur in the etalon, which will be discussed in Section 4.

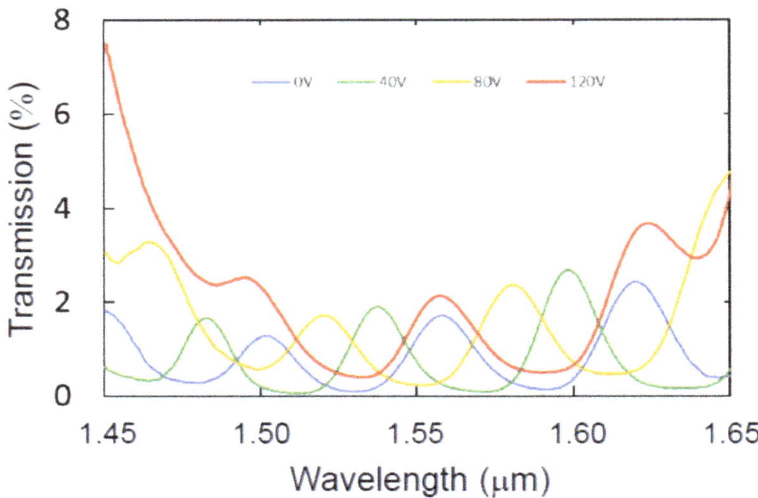

Figure 6. Transmission spectra of the developed wavelength selective filter.

Peak-transmission wavelength dependence on the applied voltage is shown in Figure 7. It is found from these results that the peak-transmission wavelength shifts to shorter wavelength as the applied voltage increases. The tuning of the resonant wavelength was expected from the Kerr effect by which refractive indexes change due to the applied voltage across the PSBP. A shift was calculated to be approximately −0.5 nm/V. The maximum shift observed was 62 nm for a 120-V voltage change and large enough for the WDM system applications. To achieve a tuning range of 7 nm to be employed in the dense WDM downlink system [4], the developed WSF can be operated at 13.5 V. The sample was capacitive and insulated electrically so that the power consumption was too small to be measured.

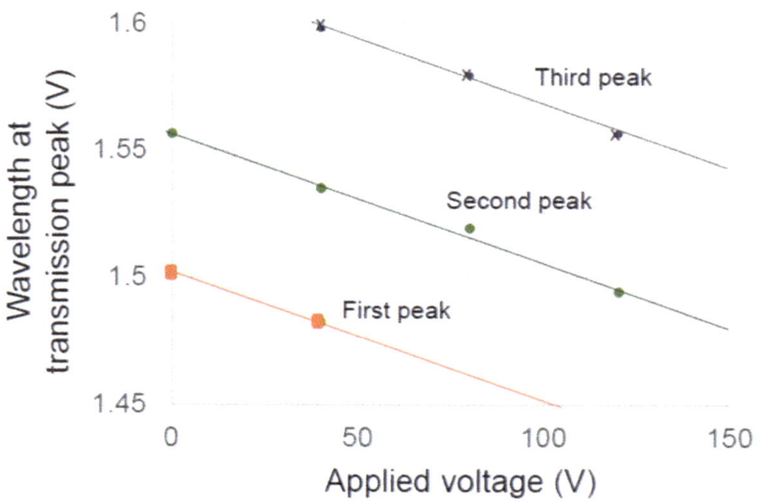

Figure 7. Wavelength at transmission versus the applied voltage. Data for three peaks are plotted.

4. Discussion

We achieved a 1.5 μm WSF with low-voltage operation and low-power consumption by using PSBP while we should analyze some data in this section. We have to make it clear of what degrades the peak transmission since a theoretical value of transmission of the absorption-free Fabry–Perot etalon is 100%. The absorption in mirror materials, silicon oxide and titan oxide, is negligible at a 1.5 mm wavelength. On the other hand, the transparent electrodes of indium tin oxide are reported to have some absorption in the telecommunication wavelength [12]. Such transparent electrodes are employed in some optical telecommunication devices although they are not used in a cavity. The light has to reflect multiplicatively inside the Fabry–Perot etalon cavity, which weakens the light intensity in general. In addition to the light absorption in the cavity and mirrors, fabrication errors, reflection at the glass substrates, and error in an incident angle are degradation factors of Fabry–Perot etalon performances.

Our current analysis model is based on the internal light absorption. We calculated the theoretical transmission spectrum (blue), as shown in Figure 8, assuming that the transparent electrodes have an extinction coefficient of 0.16. Characteristic matrix was modeled for each layer of the dielectric mirrors and transparent electrodes [13–15]. The calculated and experimental spectra (orange) exhibit similar curves with coincidence of low transmission and wide bandwidth. Our model or the calculated spectrum is traced close to the experimental spectrum. The internal light absorption in the transparent electrodes may be one of the candidates to explain the transmission attenuation. Our future research will focus on the improvement of this model by placing the electrodes outside of the cavity. Our preliminary analysis suggests that the maximum transmission will be improved.

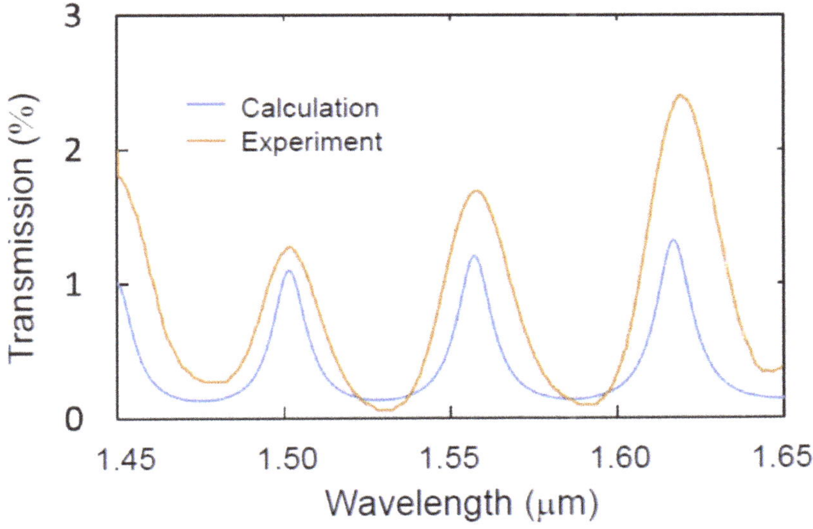

Figure 8. Transmission spectra of the experimental and calculated results with no applied voltage.

5. Conclusions

The use of polymer-stabilized blue phase (PSBP) was investigated to develop a voltage-controlled wavelength selective filter for WDM optical fiber network. It was found that a tolane-type liquid crystal introduction can increase both the blue-phase temperature range and the Kerr coefficient. A Fabry–Perot etalon filled with PSBP functioned as the wavelength selective filter, as expected. The tuning wavelength range was 62 nm at a 1.5 μm wavelength with the applied voltage from 0 to 120 V. The peak transmission was not as high as expected; however, we found the intensity attenuation may come from absorption in the transparent electrodes. We could improve the transmission with a newer structure, where the transparent electrodes are to be placed outside the cavity.

Author Contributions: S.F. carried out most of the optical measurements and wrote the manuscript. K.T. analyzed Fabry–Perot etalon characteristics. T.M., H.H. and Y.O. prepared the sample cells and made some measurements. H.K. supervised the project, especially at the material study. H.T. proposed the device configuration and fabricated the Fabry–Perot etalon substrate with transparent electrodes. All authors have read and agreed to the published version of the manuscript.

Funding: This research was funded by the Cooperative Research Program of Network Joint Research Center for Materials and Devices, grant numbers 20191316, 20201325, and 20211364.

Data Availability Statement: After publication, the data will be uploaded to https://www.eee.kagoshima-u.ac.jp/~fuku-lab/.

Acknowledgments: The authors would like to thank Fukutaro Yonekura for his technical assistance and Toshio Watanabe for fruitful discussion on telecommunication progress.

Conflicts of Interest: The authors declare no conflict of interest.

References

1. Basch, E.B.; Egorov, R.; Gringeri, S.; Elby, S. Architectural tradeoffs for reconfigurable dense wavelength-division multiplexing systems. *IEEE J. Sel. Top. Quantum Electron.* **2006**, *12*, 615–626. [CrossRef]
2. Gringeri, S.; Basch, B.; Shkia, V.; Egorov, R.; Xia, T.J. Flexible architecture for optical transport nodes and networks. *IEEE Commun. Mag.* **2010**, *48*, 40–50. [CrossRef]
3. Tasaki, K.; Tokumaru, M.; Watanabe, T.; Nagayama, T.; Fukushima, S. Nested Mach-Zender interferometer optical switch with phase generating couplers. *Jpn. J. Appl. Phys.* **2020**, *59*, SOOB04. [CrossRef]
4. Hattori, K.; Nakagawa, M.; Katayama, M.; Kani, J. Passive optical metro network based on NG-PON2 system to support cloud edges. *IEICE Trans. Commun.* **2019**, *E102B*, 88–96. [CrossRef]
5. Lin, L.Y.; Shen, J.L.; Lee, S.S.; Wu, M.C.; Sergent, A.M. Tunable three-dimensional solid Fabry-Perot etalons fabricated by surface-micromachining. *IEEE Photon. Technol. Lett.* **1996**, *8*, 101–103. [CrossRef]
6. Domash, L.; Wu, M.; Nemchuk, N.; Ma, E. Tunable and switchable multiple-cavity thin film filters. *J. Lightwave Technol.* **2004**, *22*, 126–138. [CrossRef]
7. Sneh, A.; Johnson, K.M. High-speed continuously tunable liquid crystal filter for WDM networks. *J. Lightwave Technol.* **1996**, *14*, 1067–1080. [CrossRef]
8. Yamazaki, H.; Fukushima, S. Holographic switch with a ferroelectric liquid-crystal spatial light modulator for a large-scale switch. *Appl. Opt.* **1995**, *34*, 8137–8143. [CrossRef] [PubMed]
9. Fukushima, S.; Ariki, K.; Yoshinaga, K.; Higuchi, H.; Kikuchi, H. Infrared extinction of a dye-doped (polymer/liquid crystal) composite film. *Crystals* **2015**, *5*, 163–171. [CrossRef]
10. Kikuchi, H.; Yokota, M.; Hisakado, Y.; Yang, H.; Kajiyama, T. Polymer-stabilized liquid crystal blue phases. *Nat. Mater.* **2002**, *1*, 64–68. [CrossRef] [PubMed]
11. Morishita, T.; Higuchi, H.; Tazawa, H.; Fukushima, S.; Okumura, Y.; Kikuchi, H. Polymer-stablized blue phases including tolane-type liquid crystal molecules for large Kerr effect. In Proceedings of the International Liquid Crystal Conference (ILCC 2018), Kyoto, Japan, 22–27 July 2018. paper PI-C2–38.
12. Hirabayashi, K.; Tsuda, H.; Kurokawa, T. Tunable liquid-crystal Fabry-Perot interferometer filter for wavelength-division multiplexing communication systems. *J. Lightwave Technol.* **1993**, *11*, 2033–2043. [CrossRef]
13. Born, M.; Wolf, E. *Principles of Optics*, 7th ed.; Cambridge University Press: Cambridge, UK, 1999.
14. Tsuda, H.; Fukushima, S.; Kurokawa, T. Optical triode operation characteristics of nonlinear etalons. *Appl. Opt.* **1991**, *30*, 5136–5142. [CrossRef] [PubMed]
15. Fukushima, S.; Kurokawa, T.; Ohno, M. Ferroelectric liquid-crystal spatial light modulator achieving bipolar image operation and cascadability. *Appl. Opt.* **1992**, *31*, 6859–6868. [CrossRef] [PubMed]

crystals

MDPI

Article

UV Durable LCOS for Laser Processing

Yasuki Sakurai [1,*], Masashi Nishitateno [1], Masahiro Ito [2] and Kohki Takatoh [2]

[1] SANTEC CORPORATION, 5823 Ohkusa-Nenjozaka, Komaki 485-0802, Japan; mnishitateno@santec-net.co.jp
[2] Department of Electrical Engineering, Faculty of Engineering, Sanyo-Onoda City University, Yamaguchi 756-0884, Japan; m-ito@rs.socu.ac.jp (M.I.); takatoh@rs.socu.ac.jp (K.T.)
* Correspondence: ysakurai@santec-net.co.jp; Tel.: +81-568-79-3535

Abstract: Liquid-Crystal-On-Silicon (LCOS) Spatial Light Modulator (SLM) is widely used as a programmable adaptive optical element in many laser processing applications with various wavelength light sources. We report UV durable liquid-crystal-on-silicon spatial light modulators for one-shot laser material processing. Newly developed LCOS consists of UV transparent materials and shows a lifetime 480 times longer than the conventional one in 9.7 W/cm^2 illumination at 355 nm. We investigated the durability of polymerization inhibitor mixed liquid crystal in order to extend its lifetime.

Keywords: liquid crystal; ultraviolet light; laser processing; liquid-crystal-on-silicon

Citation: Sakurai, Y.; Nishitateno, M.; Ito, M.; Takatoh, K. UV Durable LCOS for Laser Processing. *Crystals* **2021**, *11*, 1047. https://doi.org/10.3390/cryst11091047

Academic Editor: Timothy D. Wilkinson

Received: 3 August 2021
Accepted: 24 August 2021
Published: 30 August 2021

1. Introduction

Because conventional laser machining systems with one-way drawing use a 2-axis galvano mirror, they need a relatively long process time for fine large-area processing. Additionally, another mechanical actuator is needed to control depth direction on an optical axis for 3D processing. On the other hand, optical pattern forming technology allows for the generation of multiple beams or an arbitral 2D image with a computer generated hologram (CGH). Thus, high throughput and flexible processing can be realized by simultaneous multi-point scanning [1,2], one-shot 2D mask illumination or vertex beam generation, as shown in Figure 1. Spatial light modulators (SLMs) based on liquid-crystal-on-silicon (LCOS) are suitable candidates for generating the 2D optical mask, which works as both an optical phase modulator and an optical intensity modulator. It is difficult to create a CGH image with conventional techniques such as thin-film-transistor liquid crystal displays or digital mirror devices (DMD) because they have no ability of optical phase modulation and can only control optical intensity. Since LCOS can generate not only digital binary ON-OFF filters but also arbitral analog filters with a desired 2D space intensity profile, an unprecedented level of laser processing can be achieved by controlling the 3D space, including depth direction, rather than just the 2D plane [3–5]. The 1064 nm or 532 nm Nd:YAG lasers are widely used for laser processing techniques, including laser micromachining, laser printing and laser modification with a variety of materials, such as metals, dielectrics, polymers and semiconductors [6,7]. In general, the operational optical bandwidth of LCOS is limited from the visible (VIS) to the near-infrared (NIR) range, owing to an ultraviolet absorption of the constitutional materials; although some effective studies for liquid crystal devices under ultraviolet (UV) light have been reported [8–11]. However, there are so many interesting applications in the UV range [12–14], such as 3D printing, integrated circuit (IC) laser trimming, printed circuit board (PCB) cutting and maskless exposure. In this paper, we report a UV durable LCOS technology for LCOS based UV laser processing.

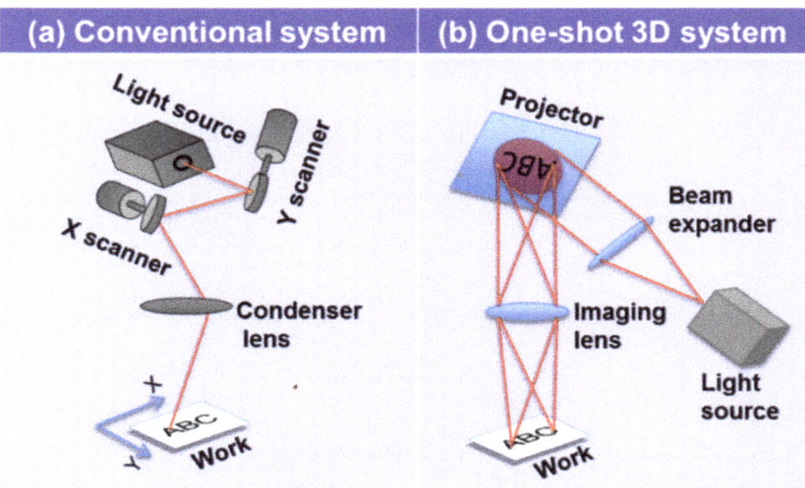

Figure 1. Schematic comparison of laser processing systems. (**a**) Conventional system with a 2-axis scanner and (**b**) one-shot 3D system.

2. Materials and Methods

Figure 2 shows a typical schematic configuration of reflective type LCOS [15–19]. The LCOS generally comprises of a liquid crystal panel and a CMOS drive circuit. The liquid crystal panel has almost the same structure as conventional liquid crystal displays, which is the sandwich-like structure consisting of cover glass plate, liquid crystal (LC) and CMOS circuit backplane. Both the cover glass and the CMOS backplane have additional functional layers to control LC orientation, as shown in Figure 2.

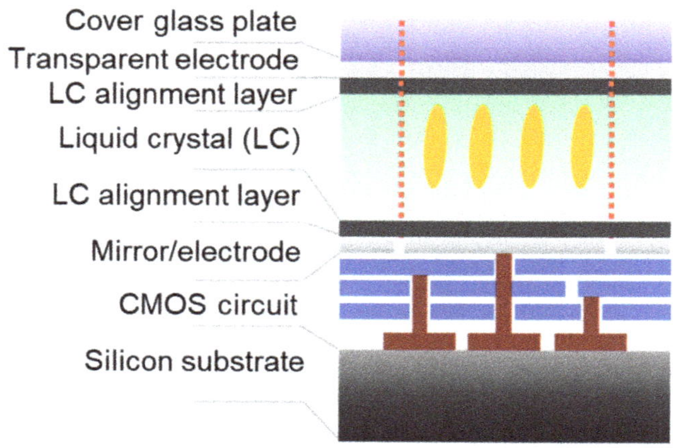

Figure 2. Schematic drawing of typical LCOS layer structure.

LCOS technology was established to obtain the basis for display applications and success in cinema projectors and rear projection TVs. Thus, typical LCOS is designed on the condition of VIS light control, and the use of LCOS has been limited in the VIS to NIR region because of the UV absorption of the constituent materials, especially in cover glass plate, indium tin oxide (ITO) electrode, polyimide LC alignment layer, metal mirror/electrode and LC material. Actually, conventional LCOS is subjected to easily permanent, irreparable damage due to UV laser radiation or high power laser radiation, even in VIS or NIR light sources [20]. In order to extend LCOS applications to the UV region, we investigated the resistivity of the LCOS composed of UV transparent materials, fused-silica cover plate, a 15 nm thin ITO electrode, an inorganic SiOx alignment layer, a SiO_2/SiN dielectric mirror on aluminum electrodes and a special blended nematic LC, which has an absorption edge wavelength (λ_g) shorter than 300 nm, as shown in Figure 3. The LC was consisted of a mixture of cyclohexane derivatives, having fluorine substitutes instead of a mixture of cyano group derivatives, which has been employed for visible applications in general. In addition, the LC contains none of the tolan derivatives, and the birefringence Δn is 0.08.

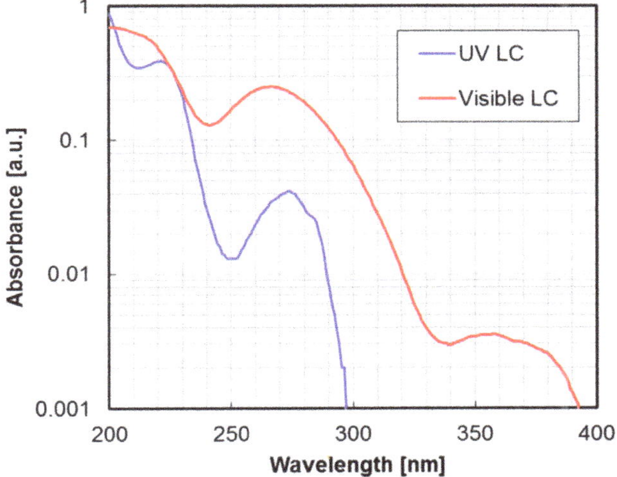

Figure 3. Comparison of the liquid crystal absorption spectrum between the typical visible liquid crystal (LC) and the special blended LC for high UV resistivity. The birefringence Δn for UV LC and visible LC was 0.08 and 0.20 at 589 nm, respectively. Each LC was dissolved in cyclohexane solvent with 20 mg/L concentration and filled into a 1 cm long silica glass cell for spectroscopic measurement.

Figure 4 shows the measurement setup for evaluating the LCOS durability of UV laser illumination. The setup consists mainly of two lasers and LCOS. Laser 1 (λ = 355 nm) is for the power laser to evaluate LCOS optical durability and Laser 2 (λ = 532 nm) is for the probing laser to optically detect LCOS damage due to UV illumination. Both laser beams are illuminated in the same spatial position onto the LCOS panel after passing a thorough wire grid polarizer to ensure that each input polarization is linear and that the axis coincides with the longitudinal axis of LC molecules for optical phase modulation. A blazed phase grating was written to the LCOS to steer the incoming laser beam to the 1st order diffraction location in the far field by a specific voltage pattern applied to the electrodes. After passing the band-pass filter to block UV lasers, only the 1st order diffracted light of 532 nm probing laser is detected for the in-situ optical monitoring of LCOS damage. For splitting 0th and 1st order diffracted light, the laser beams were irradiated at 5 degrees to the LCOS.

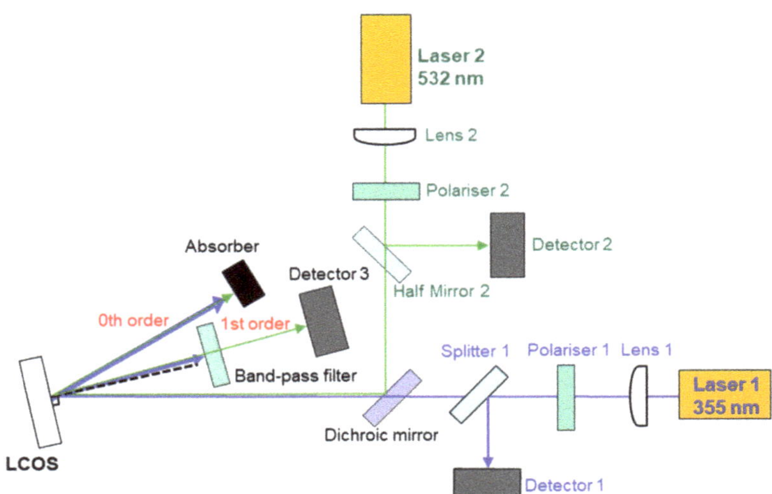

Figure 4. Optical setup for evaluating the UV laser durability of LCOS.

3. Results and Discussion

Figure 5 shows the comparison of UV resistivity between typical visible LCOS and UV LCOS. The power density and beam diameter of the irradiated UV laser was 9.7 W/cm^2 and 6 mm, respectively. In visible LCOS, an abnormality in visual appearance was observed in an irradiation time of 2 min. On the other hand, similar abnormality was observed after 16 hours in UV-LCOS, which corresponds to 480 times higher resistivity.

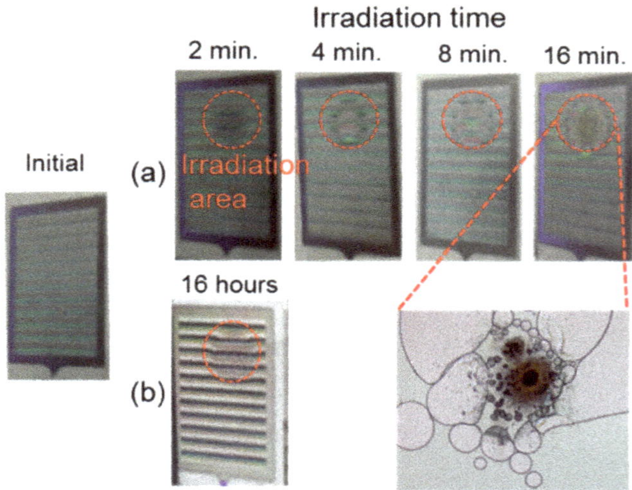

Figure 5. UV resistivity of LCOS with (**a**) typical visible LC and (**b**) special blended LC for high UV resistivity. Each cell gap of the LCOS panels for visible LC and UV durable LC is 9.7 μm and 6.0 μm, respectively.

We also measured the 1st order diffraction efficiency on a UV illuminated area, and the phase change due to UV irradiation was estimated, as shown in Figure 6. The optical phase change is calculated from measured diffraction efficiency by the following Equation (1), where η, Φ(x), Λ and m are the relative diffraction efficiency, phase shift function, grating period, and diffraction order number, respectively.

$$\eta = \left| \frac{1}{\Lambda} \int_0^\Lambda \exp(j\Phi(x)) \times \exp(-j\frac{2 \times \pi \times m \times x}{\Lambda}) dx \right|^2 \quad (1)$$

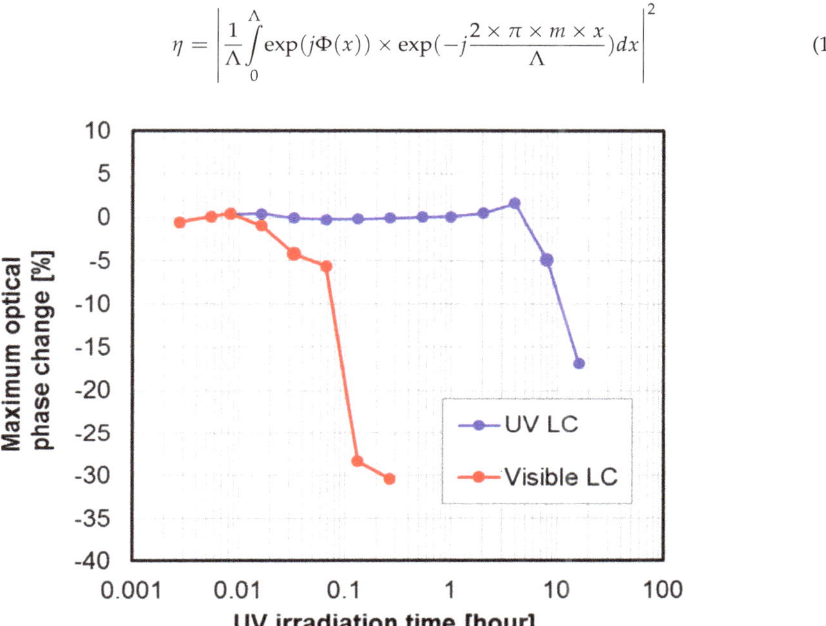

Figure 6. Maximum phase retardation on the UV irradiated area of LCOS with typical visible LC and special blended LC for high UV resistivity.

In LCOS, the maximum optical phase shift is limited by a LC thickness implicating optical response speed and fringing field effect [21–23]. Thus, a blazed grating for optical beam steering is represented by a modulo-2π (2π folded) sawtooth phase profile, and the phase shift function is expressed as a quantized multilevel step function due to the spatially pixelized mirror structure of LCOS [24–26]. Although a drastic improvement in lifetime was confirmed in UV LCOS, the phase degradation of about 15% in the irradiation time of 16 h was observed. LC is the only organic substance within UV LCOS materials on the light path and has the lowest resistance. Thus, we speculate that the observed degradation results from LC damage by UV radiation. For finding the true root cause of the damage, material content analysis by Gas Chromatography-Mass Spectrometry (GC-MS) (Agilent Technologies, Inc., Santa Clara, CA, USA) analysis was conducted, and the impurities contained in normal LC and damaged LC due to UV irradiation were compared, as shown in Figure 7. Due to its defluorinated compound close to the signal peak from the original chemical compound, multiple signals were detected from the damaged LC. The peak (1) is the one for the compound produced by the UV radiation in the damaged liquid crystal. The compound is considered to be formed by extracting fluoro atom from the original compound represented by the peak (2). Similarly, the peaks (3) and (5) are considered to be the ones for the compounds formed by the fluoro atom extraction from the original compounds shown by the peak (4) and (6). These results show that the damaged liquid crystal contains the compounds formed by fluoro atom extraction from the original compounds.

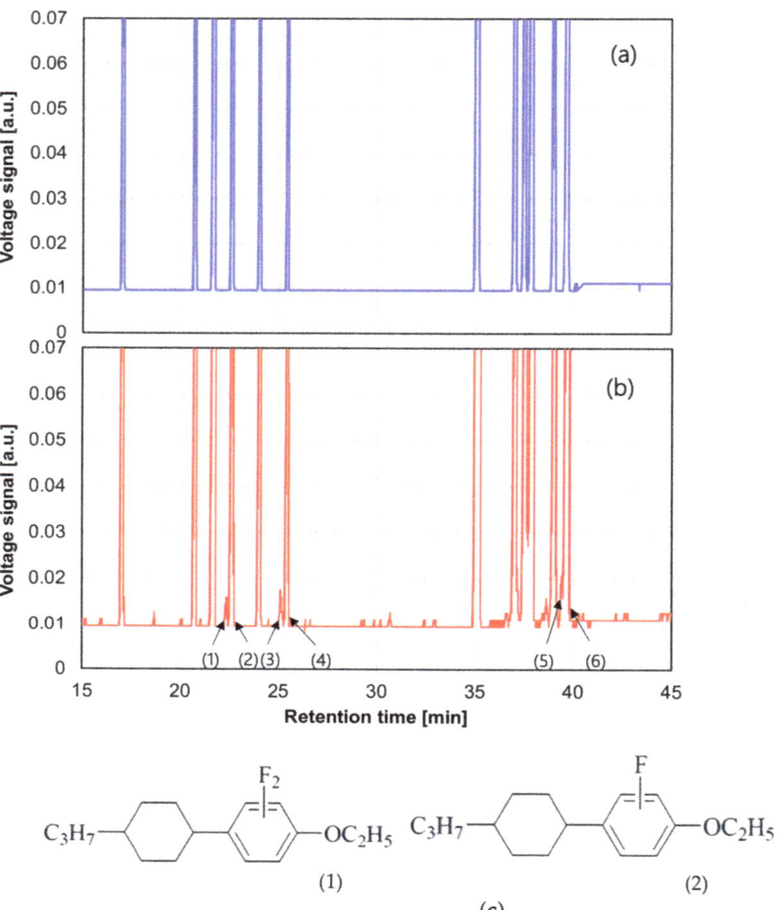

Figure 7. Measured GC/MS spectra. (**a**) normal liquid crystal without UV irradiation and (**b**) damaged liquid crystal with UV irradiation. (1)(2), (3)(4) and (5)(6) are pairs of the original base compound of the liquid crystal and its defluorinated compound, respectively. (**c**) shows a putative compound structure of (1) and (2) in the GC/MS spectrum.

It is considered that the fluorine extracted compounds were formed by the cleavage of the C-F bond and radical formation. These results should also suggest the possibility that the formed radical species could decompose the liquid crystal compounds and induce the polymerization reaction. In order to verify the assumption, the laser durability of the LC blended phenolic polymerization inhibitor (2,6-Di-*tert*-butyl-*p*-cresol) was investigated with various mixed ratio, as shown in Figure 8. The power density and beam diameter of the irradiated CW laser (λ_c = 450 nm) are 30 W/cm^2 and 1 mm, respectively. The lifetime was determined by the time when permanent visible damage was observed. With an increase of inhibitor concentration, the lifetime was extended drastically, although the nematic-isotropic phase transition temperature (T_{NI}) decreased linearly. In addition, a significant change in the viscosity of the mixed LC was not observed in the experiment range. A 30-times longer lifetime was confirmed in the LC of 6 wt% inhibitor concentration compared to the normal LC without the inhibitor. At this moment, it is impossible to identify which LC mixtures in Figure 7 were stabilized by inducing polymerization inhibitors, but it might be clarified after GC-MS analysis of the damaged LC and the comparison of the damaged LC without polymerization inhibitors. It is expected that the polymerization inhibitor mixing technique is an effective means to expand the lifetime, even in UV light operation, and that it achieves further improvement by the deployment of a more effective polymerization inhibitor with optimum concentration.

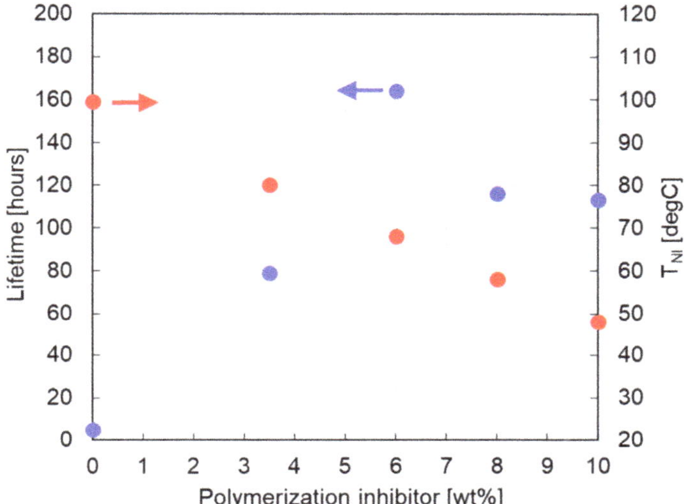

Figure 8. Measured lifetime and nematic-isotropic phase transition temperature of liquid crystal blended with various concentrations of polymerization inhibitors. The lifetime is defined as the time between the laser power turning on and visible permanent damage.

4. Conclusions

We reported UV durable LCOS structures for one-shot laser processing. We observed higher UV resistivity in a UV LCOS composed of UV transparent materials compared to a typical visible LCOS. In addition, we showed that special blended liquid crystal with a polymerization inhibitor has the potential to drastically extend LCOS lifetime under UV operation. We believe that the proposed concept will open up a new class of laser processing.

Author Contributions: Methodology, Y.S., M.N., M.I. and K.T.; writing—original draft preparation, Y.S.; writing—review and editing, K.T.; supervision, K.T. All authors have read and agreed to the published version of the manuscript.

Funding: This research received no external funding.

Conflicts of Interest: The authors declare no conflict of interest.

References

1. Hayasaki, Y.; Sugimoto, T.; Takita, A.; Nihsida, N. N. Variable holographic femtosecond laser processing by use of a spatial light modulator. *Appl. Phys. Lett.* **2005**, *87*, 031101. [CrossRef]
2. Matsumoto, N.; Itoh, H.; Inoue, T.; Otsu, T.; Toyoda, H. Stable and flexible multiple spot pattern generation using LCOS spatial light modulator. *Opt. Express* **2014**, *22*, 24723–24733. [CrossRef] [PubMed]
3. Hasegawa, S.; Hayasaki, Y. Dynamic control of spatial wavelength dispersion in holographic femtosecond laser processing. *Opt. Lett.* **2014**, *39*, 478–481. [CrossRef]
4. Matthieu, B.; Arnaud, R.; Bruno, B.; Kevin, B.; Mona, T.; Thierry, C.; Martin, R.; Lionel, C. Beat the diffraction limit in 3D direct laser writing in photosensitive glass. *Opt. Express* **2009**, *17*, 10304–10318.

5. Gian-Luca, R.; Cemal, E.; Ralf, H. Femtosecond laser direct generation of 3D-microfluidic channels inside bulk PMMA. *Opt. Express* **2017**, *25*, 18442–18450.
6. Sakurai, Y.; Hotta, Y.; Ottowa, R.; Nishitateno, M.; Zheng, L.; Yamamoto, H.; Taira, T. One-shot 3D giant-pulse micro-laser processing by LCOS direct control. Joint Session LIC+PLD+SLPC. In Proceedings of the 6th Laser Ignition Conference 2018, OPTICS & PHOTONICS International Congress 2018, Yokohama, Japan, 23–27 April 2018.
7. Bhandari, R.; Taira, T. >6 MW peak power at 532 nm from passively Q-switched ND:YAG/Cr^{4+}:YAG microchip laser. *Opt. Express* **2013**, *19*, 19135–19141. [CrossRef] [PubMed]
8. Wu, S.T.; Ramos, E. Polarized UV spectroscopy of conjugated liquid crystals. *J. Appl. Phys.* **1990**, *68*, 78–85. [CrossRef]
9. Wu, S.T. Absorption measurements of liquid crystals in the ultraviolet, visible, and infrared. *J. Appl. Phys.* **1998**, *84*, 4462–4465. [CrossRef]
10. Wen, C.H.; Gauza, S.; Wu, S.T. Photostability of liquid crystals and alignment layers. *J. Soc. Inf. Disp.* **2005**, *13*, 805–811. [CrossRef]
11. Yang, Q.; Zou. J.; Li, Y.; Wu, S.T. Fast-response liquid crystal phase modulators with an excellent photostability. *Crystals* **2020**, *10*, 765. [CrossRef]
12. Hansen, W.W.; Janson, S.W.; Helvajian, H. Direct-write UV laser microfabrication of 3D structures in lithium aluminosilicate glass. In *Laser Applications in Microelectronic and Optoelectronic Manufacturing II*; International Society for Optics and Photonics: Bellingham, WA, USA, 1997; Volume 2991, pp. 104–112.
13. Antoszewski, B.; Tofil, S.; Mulczyk, K. The efficiency of UV picosecond laser processing in the shaping of surface on elastomers. *Polymers* **2020**, *12*, 2041. [CrossRef] [PubMed]
14. Ligon, S.C.; Blugan, G.; Kuebler, J. Pulsed UV laser processing of carbosilane and silazane polymers. *Materials* **2019**, *12*, 372. [CrossRef] [PubMed]
15. Zhang, Z.; You, Z.; Chu, D. Fundamentals pf phase-only liquid crystal on silicon (LCOS) devices. *Light Sci. Appl.* **2014**, *3*, e213. [CrossRef]
16. Bleha, W.P.; Lei, L.A. Advances in liquid crystal on silicon (LCoS) spatial light modulator technology. In *Display Technologies and Applications for Defense, Security, and Avionics VII*; International Society for Optics and Photonics: Bellingham, WA, USA, 2013; Volume 8736, p. 87360A.
17. Bauchert, K.; Serati, S.; Furman, A. Advances in liquid crystal spatial light modulators. In *Optical Pattern Recognition XIII*; International Society for Optics and Photonics: Bellingham, WA, USA, 2002; Volume 4734, pp. 35–43.
18. Vettese, D. Liquid crystal on silicon. *Nat. Photonics* **2010**, *4*, 752–754. [CrossRef]
19. McManamon, P.F.; Watson, E.A.; Dorschner, T.A.; Barners, L.J. Applications look at the use of liquid crystal writable gratings for steering passive radiation. *Opt. Eng.* **1993**, *32*, 2657–2664. [CrossRef]
20. Carbajo, S.; Bauchert, K. Power handling for LCoS spatial light modulators. In *Laser Resonators, Microresonators, and Beam Control XX*; International Society for Optics and Photonics: Bellingham, WA, USA, 2018; Volume 10518, p. 105181R.
21. Kelly, J. Application of liquid crystal technology to telecommunication devices. In *National Fiber Optic Engineers Conference*; Optical Society of America: Washington, DC, USA, 2006; p. NThE1.
22. Apter, B.; Efron, U.; BahatTreidel, E. On the fringing-field effect in liquid-crystal beam-steering devices. *Appl. Opt.* **2004**, *43*, 11–19. [CrossRef] [PubMed]
23. FanChiang, K.H.; Wu, S.T.; Chen, S.H. Fringing-field effects on high-resolution liquid crystal microdisplays. *J. Disp. Technol.* **2005**, *1*, 304–313. [CrossRef]
24. Serati, S.; Stockley, J. Advanced liquid crystal on silicon optical phase arrays. In Proceedings of the IEEE Conference on Aerospace, Big Sky, MT, USA, 9–16 March 2002.
25. Lensina, A.C.; Goodwill, D.; Bernier, E.; Ramunno, L.; Berini, P. On the performance of optical phased array technology for beam steering: Effect of pixel limitations. *Opt. Express* **2020**, *21*, 31637–31657.
26. McManamon, P.F.; Dorschner, T.A.; Corkum, D.L.; Friedman, L.J.; Hobbs, D.S.; Holz, M.; Liberman, S.; Nguyen, H.Q.; Resler, D.P.; Sharp, R.C.; et al. Optical phase array technology. *Proc. IEEE* **1996**, *84*, 268–298. [CrossRef]

Article

Optically Tunable Terahertz Metasurfaces Using Liquid Crystal Cells Coated with Photoalignment Layers

Yi-Hong Shih [1], Xin-Yu Lin [2], Harry Miyosi Silalahi [2], Chia-Rong Lee [1,*] and Chia-Yi Huang [2,*]

1 Department of Photonics, National Cheng Kung University, Tainan 701, Taiwan; l78071033@ncku.edu.tw
2 Department of Applied Physics, Tunghai University, Taichung 407, Taiwan; s07210046@thu.edu.tw (X.-Y.L.); harry.miyosi.s@mail.ugm.ac.id (H.M.S.)
* Correspondence: crlee@mail.ncku.edu.tw (C.-R.L.); chiayihuang@thu.edu.tw (C.-Y.H.)

Abstract: An optically tunable terahertz filter was fabricated using a metasurface-imbedded liquid crystal (LC) cell with photoalignment layers in this work. The LC director in the cell is aligned by a pump beam and makes angles θ of 0, 30, 60 and 90° with respect to the gaps of the split-ring resonators (SRRs) of the metasurface under various polarized directions of the pump beam. Experimental results display that the resonance frequency of the metasurface in the cell increases with an increase in θ, and the cell has a frequency tuning region of 15 GHz. Simulated results reveal that the increase in the resonance frequency arises from the birefringence of the LC, and the LC has a birefringence of 0.13 in the terahertz region. The resonance frequency of the metasurface is shifted using the pump beam, so the metasurface-imbedded LC cell with the photoalignment layers is an optically tunable terahertz filter. The optically tunable terahertz filter is promising for applications in terahertz telecommunication, biosensing and terahertz imaging.

Keywords: liquid crystals; metasurface; terahertz telecommunication

Citation: Shih, Y.-H.; Lin, X.-Y.; Silalahi, H.M.; Lee, C.-R.; Huang, C.-Y. Optically Tunable Terahertz Metasurfaces Using Liquid Crystal Cells Coated with Photoalignment Layers. *Crystals* **2021**, *11*, 1100. https://doi.org/10.3390/cryst11091100

Academic Editors: Kohki Takatoh, Jun Xu and Akihiko Mochizuki

Received: 18 August 2021
Accepted: 8 September 2021
Published: 10 September 2021

1. Introduction

Photoalignment has attracted much attention due to its controllability on the orientations of liquid crystals (LCs) [1–6]. Photoalignment can be achieved by doping methyl red dyes into LCs [1–3] or coating SD1 dyes onto the substrates of LC cells [4–6]. The methyl red molecules in a dye-doped LC cell sequentially undergo absorption, photoisomerization, diffusion, desorption and adsorption on the irradiated surface of the cell after the cell is irradiated with a linearly polarized green light. The competition between the desorption and the adsorption determines the orientation of the LC director in the cell. The LC director is parallel (perpendicular) to the polarized direction of the light as the desorption (adsorption) prevails over the adsorption (desorption) at a high (low) intensity of the light [1]. Methyl red dyes will be bleached by ambient light because their absorption is in green-light wavelengths. Therefore, the reliability of methyl red dyes hinders the development of methyl red dye doped LC cells. The SD1 molecules in an LC cell that is coated with SD1 dye layers sequentially undergo absorption and photoisomerization on the irradiated surface of the cell after the cell is irradiated with a linearly polarized UV light [4–6]. The LC director in the cell is perpendicular to the polarized direction of the UV light following the photoisomerization of the SD1 molecules and can be reoriented by changing the polarized direction [4–6]. SD1 dyes are more reliable than methyl red dyes because a UV component has a smaller intensity than a green component in ambient light. LC cells that are coated with SD1 dye layers have been used to develop optically controllable optical devices such as rewritable E-papers, polarization rotators, switchable optical gratings and variable optical attenuators [5,6].

Metasurfaces, which consist of arrays of split-ring resonators (SRRs), have been used to manipulate the phases, intensities and frequencies of incident lights due to their sensitivities to the refractive indices of the media that surround them [3,7–12]. The resonance frequencies of metasurfaces can be tuned using liquid crystals (LCs) because their refractive indices are changed by voltages, heat and magnetic fields. Wu et al. reported a review of tunable metasurfaces with anisotropic LCs and showed that LC-tunable metasurfaces have potential in developing advanced optical devices [13].

Chen et al. fabricated a terahertz metasurface that is imbedded into a dual-frequency LC cell. The switch of the frequency of a specific voltage tunes the resonance frequency of the terahertz metasurface by a frequency tuning range of 15 GHz. Therefore, the dual-frequency LC cell with the terahertz metasurface is an electrically controllable filter. Liu et al. deposited an LC layer on a terahertz metasurface [11]. The resonance frequency of the terahertz metasurface is tuned by heating the LC because its refractive index is changed at the heating of a thermal stage. As a result, the terahertz metasurface that is deposited with the LC layer is a thermally controllable filter. Zhang et al. made an omega-type metasurface that is imbedded into an LC cell [12]. The resonance frequency of the omega-type metasurface can be tuned by applying a magnetic field to the LC, and the omega-type metasurface has a frequency tuning range of 0.22 GHz. Therefore, the omega-type metasurface is a magnetically controllable filter. Optically tunable terahertz filters based on metasurfaces that are imbedded into LC cells have not yet been developed, and exhibit many advantages such as free electrode-patterning, ease addressability and high potential for remote control in various ambient environments. Therefore, it is of great interest to researchers to develop optically tunable terahertz filters using metasurfaces that are imbedded into LC cells. Some latest works reported that LCs can be used to fabricate advanced optical devices, such as lenses [14], beam steerer [15] and antennas [16].

This work presents the fabrication of a terahertz metasurface that is imbedded into an LC cell with photoalignment layers. The irradiation of a pump light with various polarized directions on the photoalignment layers causes the director reorientation, tuning the resonance frequency of the terahertz metasurface by a frequency tuning range of 15 GHz. A simulation reveals that the tuning of the resonance frequency is caused by the birefringence of the LC, and the LC has a birefringence of 0.13 in the terahertz region. The terahertz metasurface that is imbedded into the LC cell with the photoalignment layers is an optically tunable terahertz filter and has potential in terahertz imaging, biosensing and terahertz telecommunication.

2. Materials and Methods

Figure 1a presents the schematic configuration of a metasurface-imbedded liquid crystal (LC) cell with photoalignment layers. The LC cell was fabricated using two 188 μm thick polyethylene terephthalate (PET) substrates, which were separated by two plastic spacers with a thickness of 100 μm. A 200 nm thick silver split-ring resonator (SRR) array was deposited on one of the PET substrates using photolithography, metal evaporation and lift-off process. Figure 1b presents the dimensions of one of the SRRs in the array. The SRR has a split gap (g), inner radius (r_i), outer radius (r_o), period in the x direction and period in the y direction of 20, 26.5, 32.5, 85 and 85 μm, respectively. One of two photoalignment layers (SD1 dye, DIC Corp, Tokyo, Japan) with a thickness of 30 nm was coated on the top substrate, and the other was coated on the metasurface of the bottom substrate. A nematic LC (HTW114300-100, Fusol Material, Tainan, Taiwan) was used in the cell. The top substrate with the dye layer, LC layer, and the bottom substrate with the dye-coated metasurface compose the metasurface-imbedded LC cell. Although the silver metasurface is easily oxidized in ambient environment, it is imbedded into the LC cell. Therefore, the metasurface was reliable in this work.

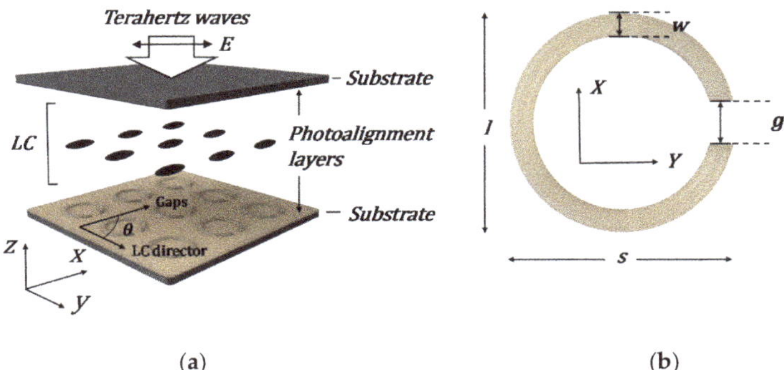

(a) (b)

Figure 1. (**a**) Schematic configuration of metasurface-imbedded LC cell with photoalignment layers. (**b**) Dimensions of SRR.

A linearly polarized pump beam (center wavelength = 365 nm) was incident to the metasurface-imbedded LC cell from the top substrate. The pump beam with an intensity of 0.3 mW/cm^2 irradiated the photoalignment layers for 100 s, aligning the LC director in the cell. The LC director made angles θ of 0, 30, 60 and 90° with respect to the gaps of the SRRs of the metasurface by changing the polarized direction of the pump beam via a polarizer. The metasurface-imbedded LC cell was placed in the chamber of a terahertz spectrometer (TPS 3000, TeraView) for studying the effect of the polarized direction of the pump beam on the resonance frequency of the metasurface. Terahertz waves that were polarized in a direction parallel to the x axis of Figure 1a were normally incident to this cell. TPS 3000 had a frequency resolution of 3 GHz in this work.

3. Results and Discussion

Figure 2 displays the experimental terahertz spectrum of the metasurface that is coated with the dye layer. This spectrum was obtained using the terahertz spectrometer in transmission mode, and the chamber in the spectrometer was filled with dry air to prevent terahertz waves from absorbing moisture. The polarization of incident terahertz waves was set parallel to the x axis of Figure 1a. The metasurface had a transmission peak in its terahertz spectrum due to the absorption of the electromagnetic resonance of the metasurface. The transmission peak was at a frequency of 0.688 THz.

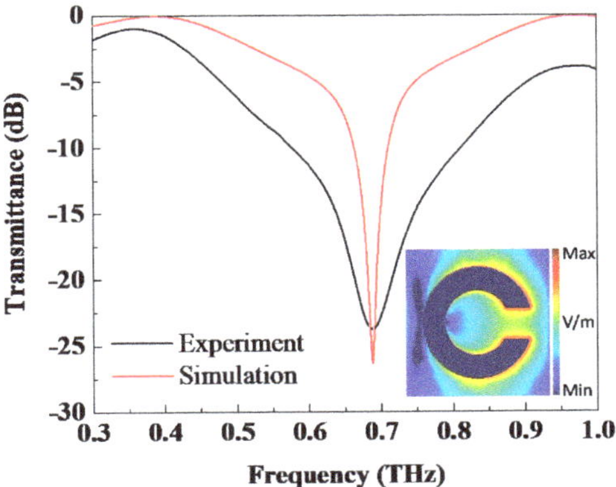

Figure 2. Experimental and simulated terahertz spectra of metasurface that is coated with dye layer. The inset presents the near-field distribution of the simulated SRR with the dye layer at 0.688 THz.

A simulation based on the finite-difference time-domain method was performed to verify the experimental spectrum of the metasurface with the dye layer. The simulated SRR had the same geometrical dimensions as the SRR of Figure 1b and was deposited on a 188 nm thick PET substrate with an area of 70 μm × 80 μm. A 30 nm thick dye layer was coated on the simulated SRR. The permittivity of the PET substrate (photoalignment) was 3.0 (1.1) in the simulation and was obtained from its time-domain spectrum. A periodical boundary condition was set in the simulation, and the conductivity of silver in the simulated SRR was 6.30×10^7 S/m. Figure 2 displays the simulated spectrum of the metasurface with the dye layer. The metasurface had a simulated peak at 0.688 THz. Therefore, the peak frequency of the simulated spectrum verifies that of the experimental spectra. The inset in Figure 2 presents the near-field distribution of the simulated SRR with the dye layer at 0.688 THz. The near field of this SRR exhibits the maximum strength at its gap. This result depicts that the electromagnetic resonance of the metasurface with the dye layer is an inductive–capacitive mode at 0.688 THz.

The SRR can be considered as an inductor–capacitor circuit, and its resonance frequency is given by [17]

$$f = \frac{1}{2\pi\sqrt{LC}} \propto \frac{1}{n}, \tag{1}$$

where L is an inductance of the inductor and C is a capacitance of the capacitor. L is determined by the enclosed area of the SRR, and C is proportional to the permittivity of a medium that is deposited on the SRR. Equation (1) depicts that the resonance frequency of the inductive–capacitive mode of a metasurface is inversely proportional to the refractive index n of a dielectric layer that is deposited on the metasurface [17]. Therefore, the resonance frequency of the metasurface with the dye layer will be sensitive to a change in the refractive index of an LC layer that is deposited on it.

Figure 3 presents the experimental terahertz spectra of an empty cell without a metasurface, empty cell with the metasurface and LC cell without a metasurface. Figure 4 displays the experimental spectra of the metasurface-imbedded LC cell at θ = 0, 30, 60 and 90°. The metasurface has resonance frequencies of 0.551, 0.557, 0.560 and 0.566 THz at θ = 0, 30, 60 and 90, respectively. The resonance frequencies of the metasurface increase with an increase in θ, and it has a frequency tuning region of 15 GHz. The increase in the resonance frequencies is caused by the birefringence of the LC. As the cell is at θ = 0° (90°), the LC director is parallel (perpendicular) to the gaps of the SRRs of the metasurface. In addition, the polarized direction of the incident terahertz waves is parallel to these gaps. Therefore, the incident terahertz waves experience an extraordinary (ordinary) refractive index of the LC in the cell at θ = 0° (90°). This result depicts that the LC has an extraordinary (ordinary) refractive index as the metasurface has a resonance frequency at 0.551 (0.566) THz. Therefore, the LC birefringence increases the resonance frequencies with the increase in θ. The results in Figure 4 reveal that the resonance frequencies of the metasurface can be tuned by the polarized directions of the pump beam. In other words, the metasurface-imbedded LC cell is an optically tunable terahertz filter. Therefore, this cell exhibits promise for applications in terahertz telecommunication, biosensing and terahertz imaging. The losses of the transmittances at frequencies that exceed 0.6 THz are caused by the multiple reflections of the metasurface-imbedded LC cell [18].

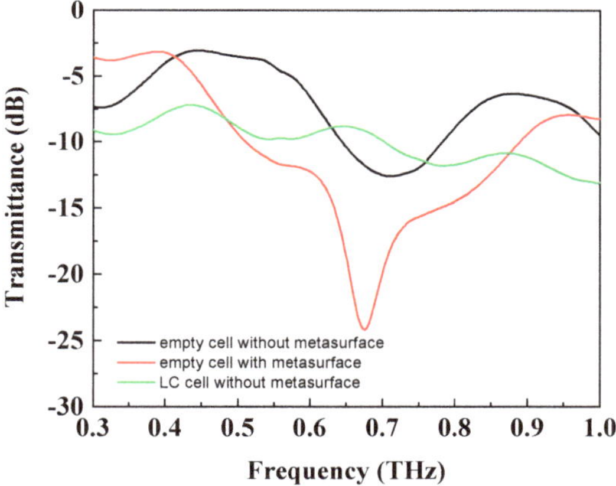

Figure 3. Experimental terahertz spectra of empty cell without a metasurface, empty cell with metasurface and LC cell without metasurface.

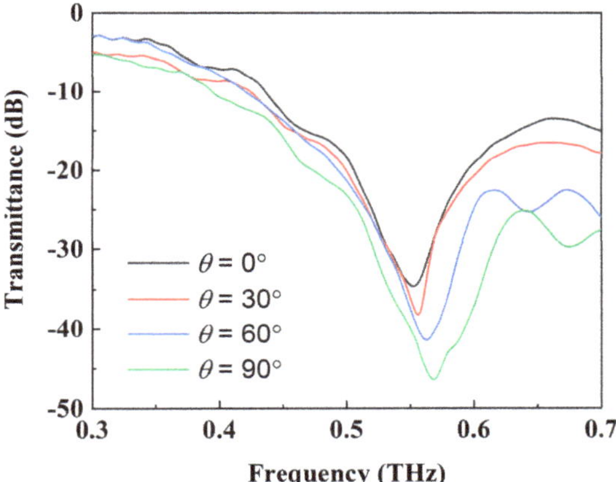

Figure 4. Experimental terahertz spectra of metasurface-imbedded LC cell at $\theta = 0, 30, 60$ and $90°$.

A simulation was performed using the software to verify the experimental spectra of the metasurface-imbedded LC cell at $\theta = 0, 30, 60$ and $90°$. A dielectric layer with a refractive index of n was deposited on the simulated dye-coated SRR (the inset of Figure 2). Figure 5 displays the terahertz spectra of the simulated dye-coated SRR with the dielectric layer at $n = 1.52$, 1.55, 1.61 and 1.65. This SRR has simulated resonance frequencies of 0.551, 0.557, 0.560 and 0.566 THz at $n = 1.65$, 1.61, 1.55 and 1.52, respectively. The resonance frequency of the simulated dye-coated SRR with the dielectric layer at $n = 1.65$ (1.52) equals that of the metasurface-imbedded LC cell at $\theta = 0°$ ($90°$). The polarized direction of the incident terahertz waves is parallel (perpendicular) to the LC director of the metasurface-imbedded LC cell at $\theta = 0°$ ($90°$), so the LC has an extraordinary (ordinary) refractive index n_e (n_o) of 1.65 (1.52) in the terahertz region. This result displays that the LC exhibits a birefringence of 0.13 in the terahertz region, and the LC birefringence increases the resonance frequencies of the metasurface as the angles between the directors and the gaps of its SRRs are increased.

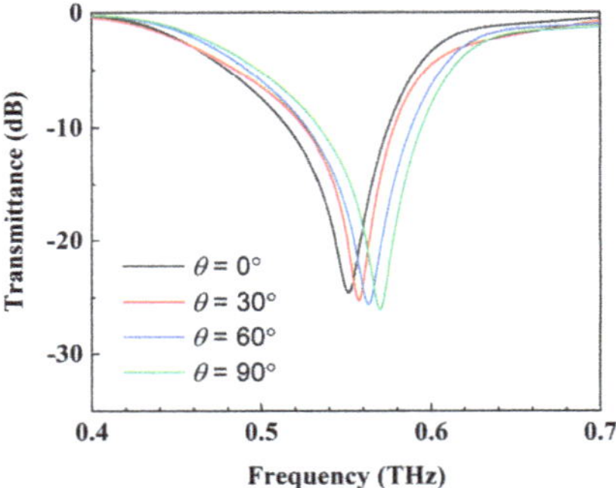

Figure 5. Simulated spectra of simulated dye-coated SRR with dielectric layer at n = 1.55, 1.62 and 1.70. n is the refractive index of a dielectric layer that is deposited on the simulated dye-coated SRR.

The effective refractive indices n_{eff_θ} of the LC in the metasurface-imbedded LC cell that involves θ = 30 and 60° can be determined by an equation of the index ellipse of the LC, and it is given by

$$n_{\mathrm{eff}\_\theta} = \sqrt{\frac{1}{\cos^2\theta/n_e^2 + \sin^2\theta/n_o^2}} \quad (2)$$

where θ is an angle between the polarized direction of the incident terahertz waves and the director of the LC. Substituting n_e = 1.65, n_o = 1.52 and θ = 30° into Equation (2) yields $n_{eff\_30^\circ}$ = 1.61. $n_{eff\_30^\circ}$ equals the refractive index of the LC in the metasurface-imbedded LC cell at θ = 30°. Substituting n_e = 1.65, n_o = 1.52 and θ = 60° into Equation (2) yields $n_{eff\_60^\circ}$ = 1.55. $n_{eff\_60^\circ}$ equals the refractive index of the LC in the metasurface-imbedded LC cell at θ = 60°. These results verify that the polarized directions of the pump beam can be used to tune the resonance frequency of the metasurface in the metasurface-imbedded LC cell.

Figure 6 is the measurement of the reliability of the metasurface-imbedded LC cell. The metasurface-imbedded LC cell was irradiated with the linearly polarized pump beam three times. The LC director made angles θ of 0, 30, 60 and 90° with respect to the gaps of the SRRs of the metasurface by changing the polarized direction of the pump beam each of the three times, and the resonance transmittances and resonance frequencies of the metasurface-imbedded LC cell at θ = 0, 30, 60 and 90° were measured by the THz spectrometer. These resonance transmittances and resonance frequencies at a given θ were the same in each of the three times. As a result, the metasurface-imbedded LC cell is reliable.

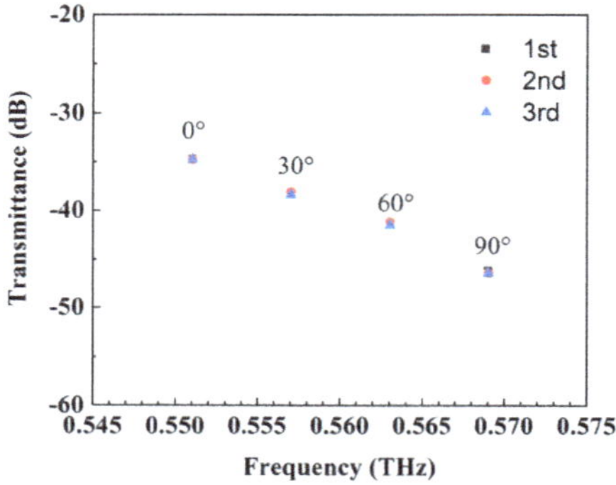

Figure 6. Measurement of reliability of metasurface-imbedded LC cell.

Atorf et al. developed an optically switchable filter using a dye-doped LC cell that is cascaded with an infrared metasurface [19]. The infrared metasurface only has two resonance frequencies before and during the irradiation of a green light. In other words, the resonance frequency of this metasurface cannot be continuously tuned by the light. Therefore, the dye-doped LC cell with the infrared metasurface is an optically switchable filter rather than an optically tunable filter. Palto et al. developed an optically switchable filter using a dye-coated visible metasurface [20]. The dye-coated visible metasurface only has two resonance frequencies at the irradiation of a blue light with two mutually orthogonal polarized directions. Therefore, the dye-coated visible metasurface is an optically switchable filter rather than an optically tunable filter. We fabricated an optically tunable filter using the dye-coated LC cell with the terahertz metasurface. The resonance frequency of this metasurface is continuously tuned by the polarized directions of the pump beam. Therefore, the dye-coated LC cell with the terahertz metasurface is an optically tunable filter. In summary, our work is different from Atorf's and Palto's works in terms of operating frequency and purpose.

4. Conclusions

This work presented the fabrication of a metasurface-imbedded LC cell with photoalignment layers. The LC director makes angles of 0, 30, 60 and 90°, and the cell has a maximum frequency tuning region of 15 GHz. The simulated results depict that the polarized direction of the pump beam reorients the LC director, tuning the resonance frequency of the metasurface. The LC has a birefringence of 0.13 in the terahertz region. The metasurface-imbedded LC cell is an optically controllable terahertz filter. The optically tunable terahertz filter can be used in biosensing, terahertz telecommunication and terahertz imaging.

Author Contributions: Conceptualization, C.-Y.H.; methodology, Y.-H.S.; software, Y.-H.S.; validation, Y.-H.S.; formal analysis, X.-Y.L., H.M.S. and C.-Y.H.; investigation, Y.-H.S., C.-R.L. and C.-Y.H.; resources, X.-Y.L., H.M.S. and C.-Y.H.; data curation, Y.-H.S. and C.-Y.H.; writing—original draft preparation, Y.-H.S. and C.-Y.H.; writing—review and editing, Y.-H.S. and C.-Y.H.; visualization, Y.-H.S. and C.-Y.H.; supervision, C.-R.L. and C.-Y.H.; project administration, C.-Y.H.; funding acquisition, C.-Y.H. All authors have read and agreed to the published version of the manuscript.

Funding: This research was funded by the Ministry of Science and Technology (MOST) of Taiwan under Contract No. MOST 110-2112-M-029-005.

Data Availability Statement: Data are contained within the article.

Conflicts of Interest: The authors declare no conflict of interest.

References

1. Ouskova, E.; Fedorenko, D.; Reznikov, Y.; Shiyanovskii, S.V.; Su, L.; West, J.L.; Kuksenok, O.V.; Francescangeli, O.; Simoni, F. Hidden photoalignment of liquid crystals in the isotropic phase. *Phys. Rev. E* **2001**, *63*, 021701. [CrossRef] [PubMed]
2. Chen, W.T.; Ji, S.C.; Chen, S.H.; Huang, C.Y.; Lo, K.Y. Competitive dye adsorption-desorption on the isotropic surface at the early stage of the photoexcitation of azo dye-doped liquid crystals. *Crystals* **2020**, *10*, 802. [CrossRef]
3. Lee, C.R.; Lin, S.H.; Wang, S.M.; Lin, J.D.; Chen, Y.S.; Hsu, M.C.; Liu, J.K.; Mo, T.S.; Huang, C.Y. Optically controllable photonic crystals and passively tunable terahertz metamaterials using dye-doped liquid crystal cells. *J. Mater. Chem. C* **2018**, *6*, 4959–4966. [CrossRef]
4. Jiang, S.A.; Sun, W.J.; Lin, S.H.; Lin, J.D.; Huang, C. Optical and electro-optic properties of polymer-stabilized blue phase liquid crystal cells with photoalignment layers. *Opt. Express* **2017**, *25*, 28179–28191. [CrossRef]
5. Chigrinov, V.; Kudreyko, A.; Guo, Q. Patterned Photoalignment in Thin Films: Physics and Applications. *Crystals* **2021**, *11*, 84. [CrossRef]
6. Chigrinov, V.; Sun, J.; Wang, X. Photoaligning and photopatterning: New LC technology. *Crystals* **2020**, *10*, 323. [CrossRef]
7. Xu, R.; Xu, X.; Yang, B.-R.; Gui, X.; Qin, Z.; Lin, Y.S. Actively Logical Modulation of MEMS-Based Terahertz Metamaterial. *Photonics Res.* **2021**, *9*, 1409–1415. [CrossRef]
8. Zhang, Y.; Lin, P.; Lin, Y.S. Tunable split-disk metamaterial absorber for sensing application. *Nanomaterials* **2021**, *11*, 598. [CrossRef] [PubMed]

9. Chiang, W.F.; Silalahi, H.M.; Chiang, Y.C.; Hsu, M.C.; Zhang, Y.S.; Liu, J.H.; Yu, Y.; Lee, C.R.; Huang, C.Y. Continuously tunable intensity modulators with large switching contrasts using liquid crystal elastomer films that are deposited with terahertz metamaterials. *Opt. Express* **2020**, *28*, 27676–27687. [CrossRef] [PubMed]
10. Chen, C.C.; Chiang, W.F.; Tsai, M.C.; Jiang, S.A.; Chang, T.H.; Wang, S.H.; Huang, C.Y. Continuously tunable and fast-response terahertz metamaterials using in-plane-switching dual-frequency liquid crystal cells. *Opt. Lett.* **2015**, *40*, 2021–2024. [CrossRef] [PubMed]
11. Liu, L.; Shadrivov, I.V.; Powell, D.A.; Raihan, M.R.; Hattori, H.T.; Decker, M.; Mironov, E.; Neshev, D.N. Temperature control of terahertz metamaterials with liquid crystals. *IEEE Trans. Terahertz Sci. Technol.* **2013**, *3*, 827–831. [CrossRef]
12. Zhang, F.; Kang, L.; Zhao, Q.; Zhou, J.; Zhao, X.; Lippens, D. Magnetically tunable left handed metamaterials by liquid crystal orientation. *Opt. Express* **2009**, *17*, 4360–4366. [CrossRef] [PubMed]
13. Xu, J.; Yang, R.; Fan, Y.; Fu, Q.; Zhang, F. A review of tunable electromagnetic metamaterials with anisotropic liquid crystals. *Front. Phys.* **2021**, *9*, 633104. [CrossRef]
14. Geday, M.A.; Caño-García, M.; Otón, J.M.; Quintana, X. Adaptive Spiral Diffractive Lenses—Lenses With a Twist. *Adv. Opt. Mater.* **2020**, *8*, 2001199. [CrossRef]
15. Zhang, Y.; Qu, X.; Li, Y.; Zhang, F. A separation method of superimposed gratings in double-projector fringe projection profilometry using a color camera. *Appl. Sci.* **2021**, *11*, 890. [CrossRef]
16. Perez-Palomino, G.; Carrasco, E.; Cano-Garcia, M.; Hervas, R.; Quintana, X.; Geday, M.A. Design and evaluation of liquid crystal-based pixels for millimeter and sub-millimeter electrically addressable spatial wave modulators. In Proceedings of the 2019 International Conference on Electromagnetics in Advanced Applications (ICEAA), Granada, Spain, 9–13 September 2019; p. 19149454.
17. O'Hara, J.F.; Singh, R.; Brener, I.; Smirnova, E.; Han, J.; Taylor, A.J.; Zhang, W. Thin-film sensing with planar terahertz metamaterials: Sensitivity and limitations. *Opt. Express* **2008**, *16*, 1786–1795. [CrossRef] [PubMed]
18. Pan, R.P.; Hsieh, C.F.; Pan, C.L.; Chen, C.Y. Temperature-dependent optical constants and birefringence of nematic liquid crystal 5CB in the terahertz frequency range. *J. Appl. Phys.* **2008**, *103*, 093523. [CrossRef]
19. Atorf, B.; Mühlenbernd, H.; Zentgraf, T.; Kitzerow, H. All-optical switching of a dye-doped liquid crystal plasmonic metasurface. *Opt. Express* **2020**, *28*, 8898–8908. [CrossRef] [PubMed]
20. Palto, S.P.; Draginda, Y.A.; Artemov, V.V.; Gorkunov, M.V. Optical control of plasmonic grating transmission by photoinduced anisotropy. *J. Opt.* **2017**, *19*, 074001. [CrossRef]

Article

Optical Filter with Large Angular Dependence of Transmittance Using Liquid Crystal Devices

Kohki Takatoh *, Masahiro Ito , Suguru Saito and Yuuta Takagi

Department of Electrical Engineering, Faculty of Engineering, Sanyo-Onoda City University, 1-1-1 Daigaku-dori, Sanyo-Onoda, Yamaguchi 756-0884, Japan; m-ito@rs.socu.ac.jp (M.I.); F216025@ed.socu.ac.jp (S.S.); F215057@ed.socu.ac.jp (Y.T.)

* Correspondence: takatoh@rs.socu.ac.jp; Tel.:+81-836-88-4544

Abstract: This study proposed a new type of optical device with variable transmittance based on the incident angle direction. These devices consist of two liquid crystal devices (LCDs) with a half-wave plate between them. Hybrid aligned nematic (HAN)-type guest-host (GH) LCDs or GH-LCDs with antiparallel alignment of high pretilt angles were used. The use of a half-wave plate allowed for the control of the p- and s-waves. Using these devices, a wide range of transmittances were obtained because no polarizer was used. The newly proposed LCDs have a wide range of applications, including use on buildings, vehicles, and glasses.

Keywords: liquid crystal; incident angle dependence; transmittance; HAN-LCD; hybrid aligned nematic; high pretilt LCD; half-wave plate; smart window; sunglasses; vehicle window

Citation: Takatoh, K.; Ito, M.; Saito, S.; Takagi, Y. Optical Filter with Large Angular Dependence of Transmittance Using Liquid Crystal Devices. *Crystals* **2021**, *11*, 1199. https://doi.org/10.3390/cryst11101199

Academic Editor: Anatoliy V. Glushchenko

Received: 7 September 2021
Accepted: 30 September 2021
Published: 3 October 2021

1. Introduction

Light control is extremely important for observing objects correctly without stress. In particular, the light observed directly from the sky to the eyes can be a large hindrance to observing an object correctly. Under typical conditions, objects are observed in the front or lower direction. Therefore, optical filters whose transmittance is small in the upward direction and large in the front or downward directions should be valuable.

Figure 1 shows the function of the optical filter with the incident angular dependence of the transmittance, which presents as a "louver" function. The louver can reduce the light only from the upward direction. Thin optical films with this function have a variety of applications, such as buildings, vehicles, and glasses. It can also be valuable if the transmittance is controlled by the electric field. This investigation used guest-host (GH) liquid crystal (LC) devices for these applications. GH-liquid crystal devices (LCDs) are devices that use dichroic dyes without polarizers. The dye molecules are aligned parallel to the molecular axis of the LC and absorb polarized light vibrating parallel to the molecular axis.

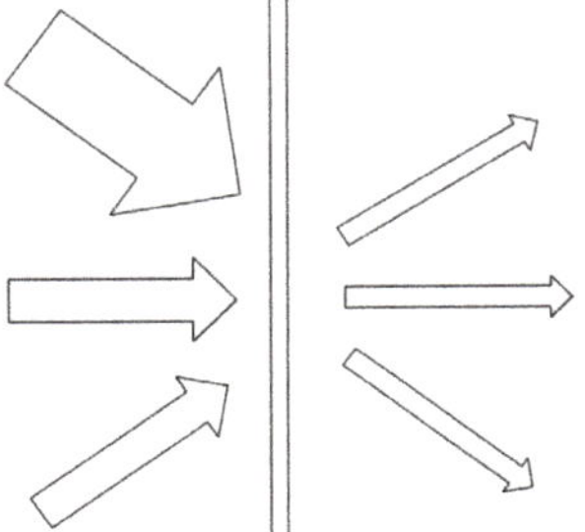

Figure 1. Optical filter of which transmittance can be controlled depending on the incident angles.

By using the structure in which the dye molecules decline to the substrate plane, the incident light propagating along the dye molecules passes through the molecules with little absorption, and the light moving perpendicular to the molecular axis is absorbed more effectively. Two types of LCDs possess this structure. The first is the hybrid aligned nematic (HAN) and the other is the high pretilt angle LCDs. High pretilt angle LCDs are defined by alignment layers showing a high pretilt angle, and the directions of LC molecular declination on both substrates are opposite. The structures of these types of LCDs are shown in Figure 2a,b.

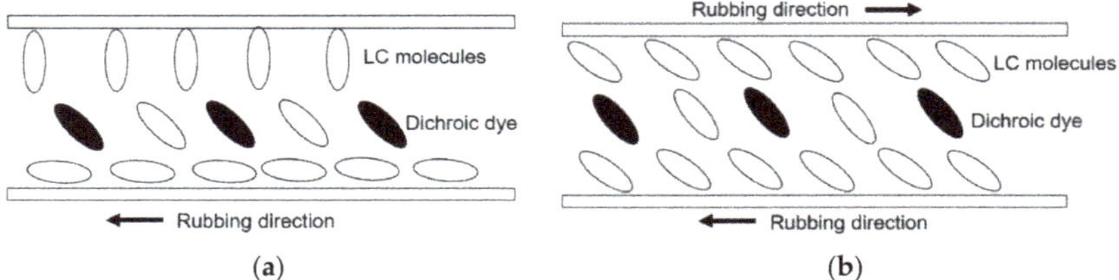

Figure 2. (**a**) Structure of HAN GH-LCDs; (**b**) Structure of high pretilt angle GH-LCDs.

As shown in Figure 2a, the two substrates of HAN GH-LCDs possess two types of alignment layers, namely a homogeneous and a homeotropic layer. The polar angle of LC molecules varies from 2 or 3 degrees (the pretilt angle of the homogeneous alignment layer) to 90° continuously from the surface of the homogeneous alignment layer to that of the homeotropic alignment layer. Therefore, the molecules at the center of the LC layer are inclined at approximately 45° from the substrate surface plane.

It was proposed that by using an inclined molecular arrangement, the incident angle dependence of the transmittance can be controlled. The reverse polymer-dispersed LC (PD-LC) mode with a HAN structure, which causes light scattering when an electric field is applied, shows the incident angle dependence of the transmittance [1–3]. This application has been proposed for smart windows as well as displays and other applications [4–7]. In these papers, except for the usage of the light scattering phenomenon [1–3], polarized light was used. This is because HAN-LCDs work only for the polarized light vibrating parallel to the LC direction. This study proposes a method for HAN-LCDs to work for non-polarized light.

Figure 2b shows the alignment layers of high pretilt angles. Alignment layers with a high pretilt angle can be realized by a mixture of the polyimide material for homogeneous alignment layers and one for homeotropic alignment layers [8].

In Figure 3, the light passes through the single HAN-LCD from the incident angles Θ and–θ, where Θ is the polar angle in the middle of the LCD. The p-wave of the incident light, from the direction Θ, vibrates parallel to the long axis of the dichroic dye and is absorbed effectively.

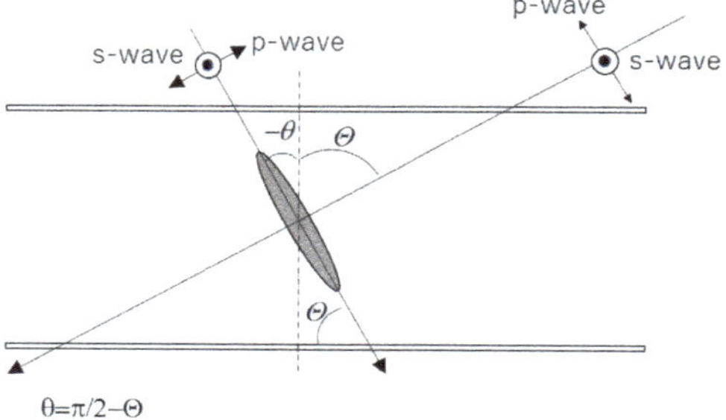

Figure 3. Light passing through the HAN or high pretilt angle LCDs from the incident angles of Θ and $-\theta$.

In contrast, the vibration of the s-wave is perpendicular to the molecular axis and the absorption is small. For the incident light from the direction of $-\theta$, both the p- and s-waves vibrate perpendicularly to the molecular axis of the dye, resulting in the absorption being small. The transmittance, even from the Θ direction, is not sufficiently reduced owing to the presence of the s-wave.

The optical device shown in Figure 4 was proposed to realize a large ratio of the light strength from the Θ and $-\theta$ directions. By utilizing this device, it is possible to realize the ratio between the largest light strength from the $-\theta$ direction and the smallest one from the Θ direction.

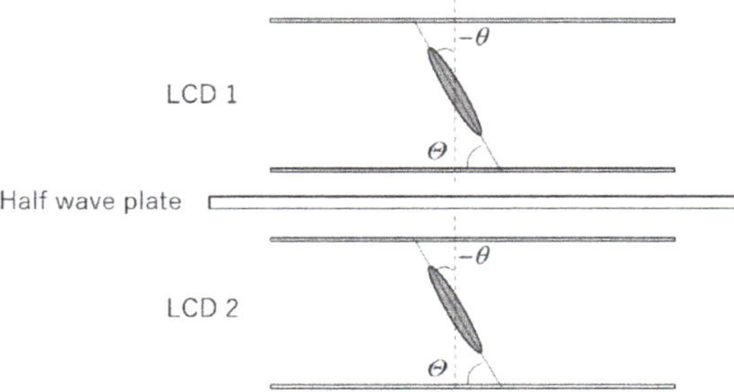

Figure 4. Structure of two-layer optically anisotropic LC filters.

Figure 5 shows the mechanism which causes the incident angle dependence of the transmittance. In the case of incident light to LCD1 from the Θ direction, the absorption of the p-wave is large, and the one of s-wave is small. By passing through the half-wave plate, the s-wave changes into a p-wave, which is effectively absorbed by passing through LCD2. However, in the case of incident light from the $-\theta$ direction, both p- and s-waves vibrate perpendicular to the dye molecular axis, passing through both LCD1 and LCD2 without large absorption. Therefore, the transmittance from the Θ direction becomes low and that from the θ direction becomes high. The device shown in Figure 4 would realize the largest ratio between the large light strength from the Θ direction and the smallest one from the $-\theta$ direction.

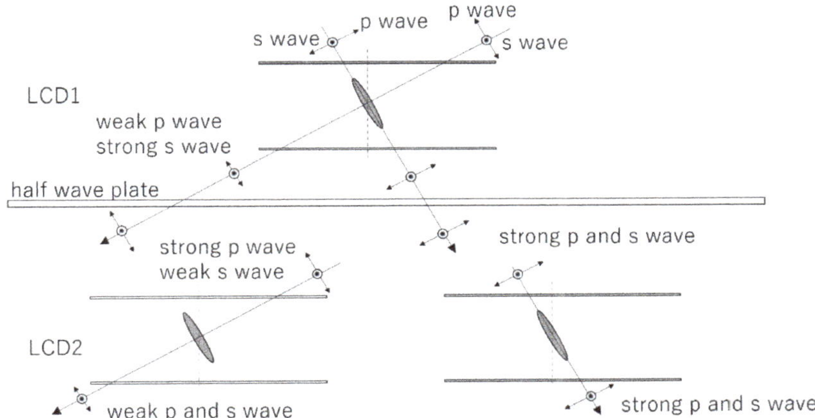

Figure 5. p-wave and s-wave through two-layer optically anisotropic LC filters in the direction of Θ and $-\theta$.

The demands for the devices which control transmittance depending on the incident angle exist in various situations. Especially, the demand to reduce only sunlight is usual.

The function of the devices described in this paper should be desired for sunglasses, sunvisors, vehicle windows, and so on. Furthermore, the demand to show information to the limited direction also exists in the fields of information displays in vehicles, displays for games, and so on. The switching function is also effective for these purposes. The films named "louver film" also have a similar function. However, those films cannot show the switching function, and the incident angle for high transmittance is limited.

2. Materials and Methods

2.1. HAN-LCD Preparations

On the surfaces of two 2 cm $\times$ 2 cm glass substrates and deposited 100 Ω/cm^2 of ITO layer, alignment layers of 100 nm were formed. For one alignment layer, the polyimide of the homeotropic alignment layer, SE4811 (Nissan Chemical Co., Tokyo, Japan), was used. For another alignment layer, a mixture of two types of polyimides, PIA-X359-01X (85%) of the homogeneous alignment layer and PIA-X768-01X (15%) of the homeotropic alignment layer (JNC Petroleum Chem. Co. Tokyo, Japan), was used. The pretilt angle of the resultant polyimide layer was 6°.

The surfaces on both alignment layers were treated by an LC alignment process called "rubbing". Utilizing an adhesive containing silica spacers of 5 μm radius, the two substrates were set, with the distance between the two alignment layers being 5 μm. The mixtures of LC material, ZLI-4792 (Merck Co. Darmstadt, Germany), and dichroic black dye NKX-4173 (Hayashibara Ltd., Okayama, Japan) or NKX-4010 (Hayashibara Ltd.) were injected at concentrations of 1, 3, and 5 wt.%.

2.2. High Pretilt Angle LCD Preparations

2.2.1. 25° Pretilt Angle

Alignment layers with high pretilt angles were formed on the surfaces of the ITO layers by using a mixture of 80 wt.% polyimide PIA-X768-01X for the homeotropic alignment layer and 20 wt.% of polyimide PIA-X359-01X for the homogeneous alignment layers. Both alignment layers were treated by rubbing. The pretilt angle of the resultant alignment layer was 25°. The two substrates with the same alignment layers were set for opposite rubbing directions, as shown in Figure 2b.

2.2.2. 40° Pretilt Angle

For the alignment layers, a mixture of 90% PIA-X768-01X and 10% PIA-X359-01X was used. The pretilt angle of the obtained alignment layer was 40°. LCDs were formed by using the same process.

2.3. Preparations of Optically Anisotropic Optical Device Using Two-Layer LCDs

For the half-wave plate, the film of Nichiban cellophane type No. 405, with a retardation of 260.4 nm (550 nm), was attached to LCD1, as shown in Figure 4. The angle between the drawing direction of the film and the rubbing direction of the LCD1 alignment film was set to 45°. On the other surface of the film, LCD2 was set as shown in Figure 4. The angle between the drawing direction of the film and the rubbing direction of LCD2 was set to 45°.

2.4. Measurements of the Pretilt Angles for the Alignment Layers

For the measurements of the alignment layer pretilt angles, LCD cells with parallel rubbing paths but opposite directions were prepared. The distance between the alignment layers was set as 20 μm. The pretilt angle was measured using the PAS-301 pretilt angle measurement system (Elscon Co., Newark, DE, USA).

2.5. Measurements of the Incident Angle Dependence of the Transmittance

The incident angle dependences of the transmittance of the LCDs and two-layer LCD devices were measured using an optical property measurement system RETS-100

(Otsuka Electronics Co., Osaka, Japan), and the relationship between the sign of the incident angles and the directions of the polar angle of the LC molecules is shown in Figure 3. All measurements were carried out by using the light of 550 nm wavelength.

3. Results and Discussions

Transmittance Dependences on the Incident Angles for HAN-LCDs

Figure 6 shows the transmittance dependence on the incident angles for HAN-LCDs using LC materials containing 1, 3, and 5 wt.% of the dichroic dye NKX-4173 (Hayashibara Ltd., Japan). In the case of 1 wt.%, the maximum value, minimum value, and ratio between them was 71%, 55%, and 1.3 respectively, in the range of ±45°.

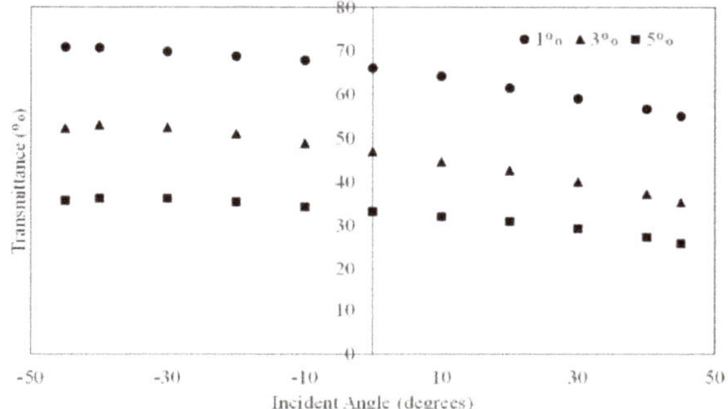

Figure 6. Transmittance dependences of the incident angles for HAN-LCDs using LC materials containing 1, 3, and 5 wt.% of NKX-4173 (Hayashibara Ltd., Japan) dichroic dye.

Figure 7 shows the dependence of the transmittance on the incident angles for the 2-layer HAN-LCDs, as shown in Figure 4, using LC materials containing 1, 3, and 5 wt.% of the dichroic dye NKX-4173 (Hayashibara Ltd., Japan). In the case of 1 wt.%, the maximum value, minimum value, and ratio was 37%, 21%, and 1.8, respectively, in the range of ±45°.

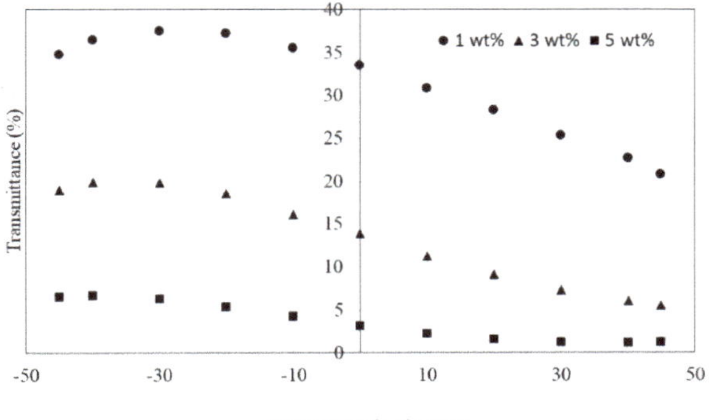

Figure 7. Transmittance dependences of the incident angles for the two-layer HAN-LCDs, as shown in Figure 4, using LC materials containing 1, 3, and 5 wt.% of dichroic dye, NKX-4173 (Hayashibara Ltd., Japan).

Figure 8 shows the behaviors of the p- and s-waves parallel and perpendicular to the dye molecules for single and two-layer HAN-LCDs. In the case of HAN-LCDs, the polar angle of the LC molecules or dye molecules is distributed from several degrees in the vicinity of one alignment layer to 90° in another. The polar angle of the LC molecule can be considered to represent either the angle in the middle of the LC layer or the average of the LC alignment.

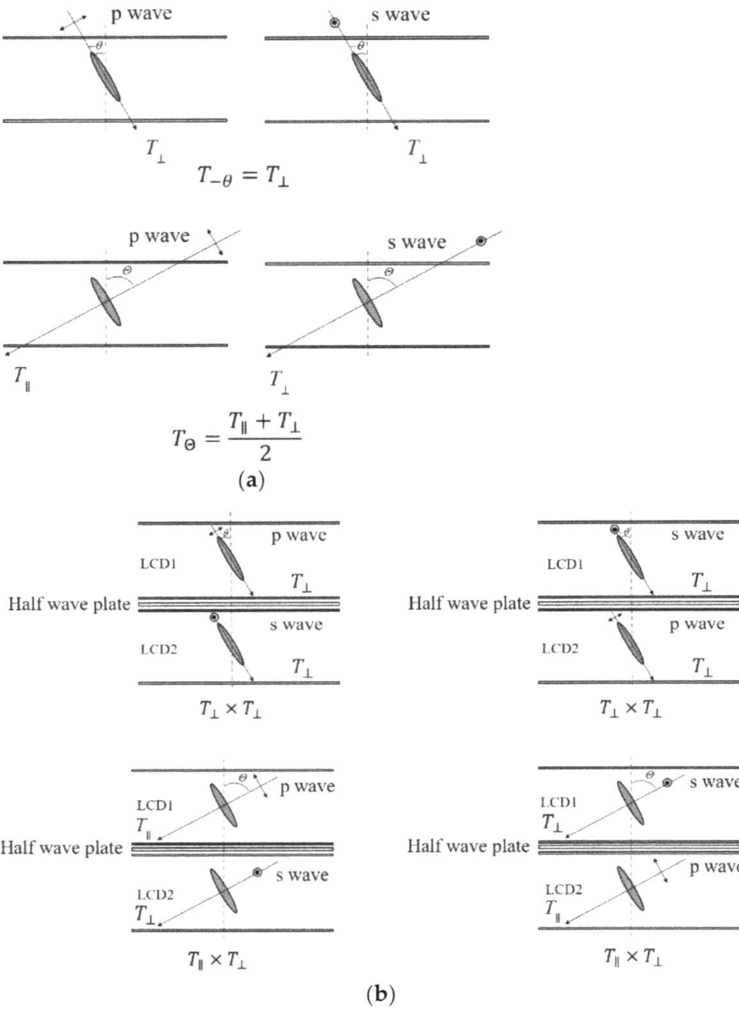

Figure 8. (**a**). Behaviors of p-waves and s-waves parallel and perpendicular to the dye molecules for single HAN-LCDs. (**b**). Behaviors of p-waves and s-waves parallel and perpendicular to the dye molecules for 2-layer HAN-LCDs.

$T_\parallel$ shows the transmittance of the polarized light vibrating parallel to the molecular axis of the dichroic dye through GH-LC materials. $T_\perp$ shows the transmittance of the polarized light vibrating perpendicular to the molecular axis of the dye.

$T_\parallel$ and $T_\perp$ are the transmittances of the p- and s-waves from the angle Θ in Figure 8a, respectively. $T_\perp$ is almost equal to the transmittance p-wave and s-wave from the angle $-\theta$. T_Θ and $T_{-\theta}$ are the transmittance of light from the incident angles Θ and $-\theta$, respectively. As shown in Figure 8a, for a single cell $T_{-\theta}$ is equal to $T_\perp$. In the case of light from the incident angle Θ, the transmittance of the p-wave is equal to $T_\parallel$ and that of the s-wave is equal to $T_\perp$. Generally, the strengths of the p- and s-waves are identical. Consequently, T_Θ can be represented by Equation (1):

$$\left(T_\parallel + T_\perp\right)/2 \tag{1}$$

and the ratio r, where $T_{\parallel} < T_{\perp}$, can be expressed as follows:

$$r = \frac{T_{-\theta}}{T_{\Theta}} = \frac{2T_{\perp}}{T_{\parallel} + T_{\perp}} \quad (2)$$

Therefore, the ratio r should be limited.

Figure 8b shows the behavior of the p- and s-waves that enter into the two-layer HAN-LCDs from the incident angle $-\theta$ and Θ. The transmittance of the p-wave through LCD1 from the incident angle $-\theta$ can be considered to be $T_{\perp}$, as described above. The p-wave changes into the s-wave by passing through the half-wave plate. The transmittance of the s-wave is also considered to be equal to $T_{\perp}$. Therefore, the transmittance of the p-wave passing through two-layer LCDs is ${T_{\perp}}^2$. The transmittance of the s-wave incident from $-\theta$ passing through two-layer HAN-LCDs is also equal to ${T_{\perp}}^2$. As a result, the transmittance of the incident light from $-\theta$ should be ${T_{\perp}}^2$.

The transmittance of the p-wave through LCD1 from the incident angle Θ can be expressed as $T_{\parallel}$. The p-wave changes to the s-wave by a half-wave plate. As a result, the p-wave transmittance from the incident angle Θ can be represented by Equation (3):

$$T_{\parallel} \times T_{\perp} \quad (3)$$

In contrast, the s-wave transmittance for LCD1 is $T_{\perp}$ and changes to a p-wave by a half-wave plate. The p-wave transmittance through LCD2 is represented by Equation (3). Consequently, the transmittance of the incident light from Θ can be expressed by Equation (3).

The ratio r of the maximum transmittance $T_{-\theta}$ and minimum transmittance T_{Θ} can be expressed as follows:

$$r = \frac{T_{-\theta}}{T_{\Theta}} = \frac{T_{\perp}}{T_{\parallel}} \quad (4)$$

For an increasing $T_{\perp}$ and $T_{\parallel}$ ratio, it is effective to increase the dichroic ratio of the dye and the order parameters of the LC materials. Increasing the dye concentration and the cell thickness is effective in increasing the r value by reducing the minimum transmittance, T_{Θ}. The required maximum transmittance is determined depending on the application. Therefore, to achieve a large r value, the dichroic ratio of the dye and the order parameters of the LC materials are important.

Figure 9 shows the dependence of the transmittance on the incident angles for HAN-LCDs using LC materials containing 1, 3, and 5 wt.% of the dichroic dye NKX-4173 (Hayashibara Ltd., Japan) with a polarizer. The optical axis of the polarizer was parallel to the alignment direction of the HAN-LCD. The transmittance was halved with the use of the polarizer. From Figure 9, it can be determined that $T_{\parallel}$ was 0.32, and $T_{\perp}$ was 0.62 for 1 wt.%. For these values, the transmittance in Fig.9 was doubled because of the usage of the polarizer. By using these values for the two-layer LCD device, $T_{-\theta} = T_{\perp}^2$ becomes 0.38 and $T_{\Theta} = T_{\parallel} \times T_{\parallel}$ becomes 0.20. These values show the good correspondence with Figure 7. By using these values, Figure 6 can also be approximately explained. In the case of 3 and 5 wt.%, the relationship can also be approximately confirmed.

The absorbances $A_{\parallel}$ and $A_{\perp}$ were measured by polarized light parallel and perpendicular to the alignment directions. Antiparallel LCD of 1° pretilt angle using ZLI-4792 and NKX-4173 was used. $A_{\parallel}$ and $A_{\perp}$ were 0.835 and 0.0725, respectively. By Lambert-Beer's law, the ratio of the absorption coefficients of dichroic dye, $\alpha_{\parallel}/\alpha_{\perp}$,was 11.5 ($\alpha_{\parallel}/\alpha_{\perp} = A_{\parallel}/A_{\perp}$). On the other hand, the transmittance of HAN LCD without a dichroic dye of 45° incident angle was 0.785. By using the value, the absorbances $A_{\parallel}$ and $A_{\perp}$ of HAN GH-LCD of 45° incident angle were calculated ($A_{\parallel} = 0.39$, $A_{\perp} = 0.102$). The ratio of absorption coefficients of 45° incident angle was 3.80. The value was reduced to be 1/3. This is considered to be the distribution of the dichroic dye polar angle from the homogeneous alignment layer to the homeotropic alignment layer.

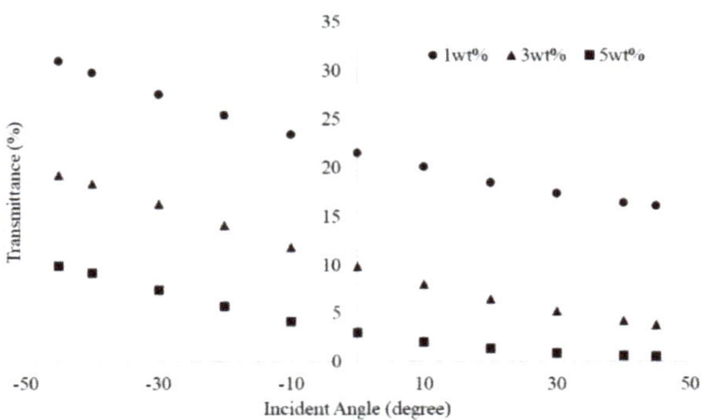

Figure 9. Transmittance dependences on the incident angles for HAN-LCDs using LC materials containing 1, 3, and 5 wt.% of NKX-4173 (Hayashibara Ltd. Japan) dichroic dye with a polarizer. The optical axis of a polarizer is parallel to the alignment direction of the HAN-LCD.

Figure 10 shows the transmittance dependencies on the incident angle for two-layer HAN-LCDs using LC materials containing 3 wt.% of dichroic dye with applied voltages of 0, 1, 3, 5, and 10 V. The same voltage was applied to the two LC panels during the measurements. By applying a voltage, the polar angle of the LC molecules increased, resulting in the asymmetric properties of the LCDs decreasing. The saturated voltage of LCDs, using the LC material ZLI-4792, was approximately 4 V. However, the asymmetric properties were still observed at applied voltages as high as 10 V because of the effect of the pretilt angle and the LC molecular arrangement maintained in the vicinity of homogeneous alignment layer.

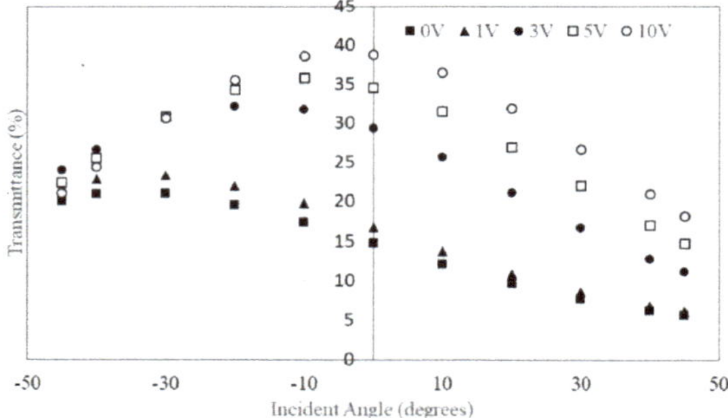

Figure 10. Incident angle dependencies of the transmittance for the two-layer HAN-LCDs using LC materials containing 3 wt.% of dichroic dye with applied voltages.

Figures 11 and 12 show the dependence of the transmittance on the incident angle for the two-layer LCDs, as shown in Figure 2b, with pretilt angles of 25° and 40°, respectively. In the case of the two-layer LCDs with a 25°pretilt angle, the maximum value was 28% at an incident angle of −30°. The value at 45° was 8% and the ratio r was found to be 3.5. In the case of two-layer LCDs with a 40° pretilt angle, the maximum value was 35% at −30°. The value at 45° was 8% and the ratio r was found to be 4.4. In the case of two-layer HAN-LCDs with 1 wt.% dye concentration, the maximum value, the value of 45 degrees and the ratio were 37%, 21% and 1.8, respectively. In the case of high pretilt angle LCDs, all the LC molecules were expected to be aligned in the same direction, although those of HAN-LCDs were distributed. As a result, the ratio of two-layer high pretilt angle LCDs was larger than that for HAN-LCDs.

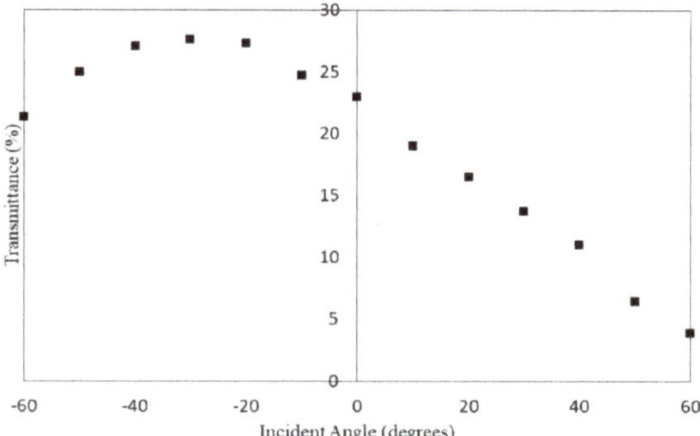

Figure 11. Transmittance dependence on the incident angle for the 2-layer LCD with a 25° pretilt angle using a concentration of 1 wt.% NKX-4010 (Hayashibara Ltd., Japan).

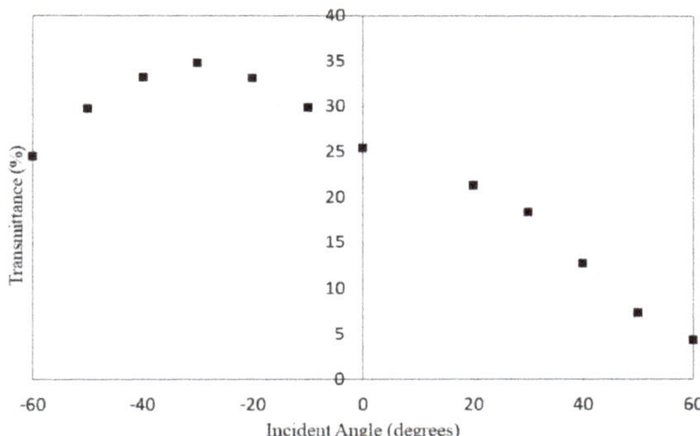

Figure 12. Transmittance dependence on the incident angle for the two-layer LCDs of 40° pretilt angle. The concentration of NKX-4010 (Hayashibara Ltd., Japan) dye was 1 wt.%.

In the case of high pretilt angle LCDs, the incident angle of the maximum value was approximately 30°. This is different from the incident angle, which was expected from the polar angle or pretilt angle. The incident angle of the maximum transmittance depends not only on the polar angle but also on the optical path length and reflection, as expected from the Fresnel law. Based on these three factors, both high pretilt two-layer LCDs were considered to show the maximum value at approximately 30°.

By using Fresnel's equations, the transmittance through glass (refractive index: 1.5) from air was calculated. When the transmittance perpendicular to the glass is 80 degrees, the transmittance for p-wave/s-wave are 77%//76% at 30 degrees and 64%/58% at 60 degrees, respectively. By increasing the incident angle, the transmittance decreases and the difference of the transmittance between p-wave and s-wave increases. In Figure 9, only the p-wave was used by a polarizer. By increasing the incident angle at (−) direction, the transmittance increases monotonically. However, in Figures 7, 11 and 12, both s-waves and p-waves were used. This should be one of the reasons why the maximum value is observed. The same tendency is also observed in Figure 6.

4. Conclusions

This study proposed a series of devices capable of different transmittances based on incident angle direction. The devices consist of either two HAN GH-LCDs or GH-LCDs with a high pretilt angle antiparallel alignment and a half-wave plate between them. Both the p- and s-waves for the incident light can be controlled, owing to the rotation of the wave vibration direction by the half-wave plate. The dependences can be controlled by the applied voltage. The ratio of the devices using GH-LCDs with high pretilt antiparallel alignment was found to be higher than that using HAN GH-LCDs. This is because in the case of HAN GH-LCDs, the polar angle at the LC materials or dichroic dye is distributed from 2 or 3 degrees to 90 degrees.

Due to the capabilities of these devices, they have a wide range of application, including buildings, vehicles, and glasses.

Author Contributions: K.T. proposed and conceptualized the devices, supervised the project, and wrote the manuscript. S.S. and Y.T. prepared the sample cells and made measurements. M.I. supervised S.S. and Y.T. and analyzed the data. All authors have read and agreed to the published version of the manuscript.

Funding: This research received no external funding.

Institutional Review Board Statement: Not applicable.

Informed Consent Statement: Not applicable.

Data Availability Statement: Not applicable.

Acknowledgments: The authors would like to extend their sincerest gratitude to Chisso Petro Chemical Co. and Nissan Chemical Co. for providing the polyimide materials used during this investigation. The authors would also like to thank Merck Co. for providing the LC materials and Hayashibara Co. for the dye materials used during this investigation.

Conflicts of Interest: The authors declare no conflict of interest.

References

1. Yamaguchi, R.; Waki, Y.; Sato, S. Wide viewing angle properties in nematic liquid crystal/UV curable liquid crystal composite film with some aligned-modes. *J. Photopol. Sci. Technol.* **1997**, *10*, 19–24. [CrossRef]
2. Yamaguchi, R.; Inoue, K.; Oikawa, Y.; Takasu, T. Electro-optical property in hybrid aligned reverse mode cell using liquid crystals with positive and negative dielectric constant anisotropies. *J. Photopol. Sci. Technol.* **2017**, *30*, 463–466. [CrossRef]
3. Yamaguchi, R.; Ushizaki, R. Louver function in hybrid aligned reverse mode using dual frequency liquid crystal. *J. Photopol. Sci. Technol.* **2019**, *32*, 545–548. [CrossRef]
4. Hong, S.H.; Jeong, Y.H.; Kim, H.Y.; Lee, S.H. Novel nematic liquid crystal device associated with alignment controlled by fringe field. *Jpn. J. Appl. Phys.* **2001**, *40*, L272. [CrossRef]
5. Kasajima, Y.; Kato, T.; Kubono, A.; Tasaka, S.; Akiyama, R. Wide viewing angle of rubbing-free hybrid twisted nematic liquid crystal displays. *Jpn. J. Appl. Phys.* **2008**, *47*, 7941–7942. [CrossRef]
6. Lim, Y.J.; Song, J.H.; Kim, Y.B.; Lee, S.H. Single gap transflective liquid crystal display with dual orientation of liquid crystal. *Jpn. J. Appl. Phys.* **2004**, *43*, L972. [CrossRef]
7. Atorf, B.; Mühlenbernd, H.; Zentgraf, T.; Kitzerow, H. All-optical switching of a dye-doped liquid crystal plasmonic metasurface. *Opt. Express* **2020**, *28*, 8898–8908. [CrossRef] [PubMed]
8. Takatoh, K.; Akimoto, M.; Kaneko, H.; Kawashima, K.; Kobayashi, S. Molecular arrangement for twisted nematic liquid crystal displays having liquid crystalline materials with opposite chiral structures (reverse twisted nematic liquid crystal displays). *J. Appl. Phys.* **2009**, *106*, 64514. [CrossRef]

Article

Stabilization of Long-Pitch Supertwisted Nematic Structures

Masahiro Ito *, Satoshi Ohmi and Kohki Takatoh

Department of Electrical Engineering, Faculty of Engineering, Sanyo-Onoda City University, Daigakudori 1-1-1, Sanyo-Onoda 756-0884, Japan; f119603@ed.socu.ac.jp (S.O.); takatoh@rs.socu.ac.jp (K.T.)
* Correspondence: m-ito@rs.socu.ac.jp

Abstract: Stabilized reverse twisted nematic liquid crystal devices (RTN-LCDs) were fabricated using formation of a polymer matrix under UV irradiation with an applied voltage (sustain voltage) in the vicinity of the alignment layers. In the absence of an applied voltage, the non-stabilized RTN structure gradually returns to a splay twist structure. The sustain voltage was decreased with an increase in temperature. A stabilized long-pitch supertwisted nematic (LPSTN) structure could also be formed during the RTN structure stabilization process with a much lower sustain voltage at a temperature near the clearing point. The chiral pitch for the LPSTN structure is longer than that for a typical STN structure. LPSTN-LCDs similar to RTN-LCDs show a large reduction in both the threshold and saturation voltage compared with those for TN-LCDs consisted of the same LC materials. Furthermore, a notable feature of LPSTN-LCDs is a change to a TN structure when a high voltage is applied. A black state can be realized due to the change from the LPSTN structure to the RTN structure unlike the typical STN mode under the crossed nicols condition. In contrast STN-LCDs retain their color due to the retardation because the RTN and LPSTN states are considered topologically equivalent.

Keywords: twisted nematic; supertwisted nematic; low driving voltage; polymer stabilization

Citation: Ito, M.; Ohmi, S.; Takatoh, K. Stabilization of Long-Pitch Supertwisted Nematic Structures. *Crystals* **2021**, *11*, 1541. https://doi.org/10.3390/cryst11121541

Academic Editors: Stephen J. Cowling, Kohki Takatoh, Jun Xu and Akihiko Mochizuki

Received: 10 November 2021
Accepted: 7 December 2021
Published: 9 December 2021

1. Introduction

A twisted hybrid aligned [1], hybrid twisted nematic (HTN) [2], hybrid aligned nematic using strong and weak polar anchoring surfaces [3], or reverse twisted nematic (RTN) [4–7] liquid crystal displays (LCDs) have been proposed to reduce the driving voltage. An HTN that uses strong and weak polar anchoring surfaces was numerically analyzed [8]. Long-pitched supertwisted nematic (LPSTN) LCDs have been reported [9] in the development of RTN-LCDs. In addition, the doping of metallic nanoparticles in the LC system [10], or ferroelectric barium titanate nanoparticles in alignment layer [11] provide low power consumption in the display devices.

The RTN and LPSTN structures were separately formed controlling the chiral pitch to the cell thickness (p/d) ratio [9]. These structures were changed from a splay twisted (ST) structure (Figure 1a), which includes a chiral reagent causing the twist distortion opposite to that determined by the rubbing directions and were formed by the application of voltage over Fredericks transition threshold voltage to the ST structure. The twist of the liquid crystal is basically determined by the angle of the director in the vicinity of the alignment layer depending on the rubbing direction of the alignment layers of the upper and lower substrates. The director of twisted nematic and STN structures, which include a chiral reagent causing the twist distortion that determined by the rubbing directions, rotates in the direction of Azimuth angle. In addition, the STN has a steeper transmittance change than the TN. On the contrary, splay-twisted nemantic structures which include a chiral reagent causing the twist distortion opposite to that determined by the rubbing directions, which rotate in the direction of Azimuth and Polar angles. The RTN and LPSTN structures that include a chiral reagent with the twist distortion opposite to that determined by the rubbing directions have a metastable state and can be maintained when a specific voltage is applied after application of the saturation voltage. This voltage is referred to as the sustain voltage. The ST structure changes to RTN or LPSTN structures as the director near the center of the LC cell is rotated in the opposite direction, as the Gibbs free energy increases with the voltage. This phenomenon can be explained by assuming that the stress of the LC structure generated by the splayed structure with an electric voltage can be released because the reversely twisted structure (no splay deformation) is more stable than the splayed structure [5,12,13]. The RTN structure appears when the p/d ratio is more than three. Moreover, our group has developed a stabilization method for the RTN structure by UV irradiation using reactive mesogens (RMs) with application from 1.25 to 1.5 V [14]. The driving voltage for the RTN-LCD is reduced compared with the TN-LCD (Figure 1). When the p/d ratio is 2.78, the STN structure appears [9]. In the case that the rubbing directions are twisted at 90° by application of high voltage to avoid ST structure, it changes into the RTN structure [5] and, at the same time, the STN structure in which LC material is twisted at 270° could be formed by twisting in the direction opposite to the RTN. These RTN and STN do not contain splayed structures. The p/d value approaches 1.5, which is necessary for the LC to twist 270°; therefore, the ST structure is formed when the voltage is 0 V. Application of a voltage higher than the threshold voltage causes the energy of the ST structure to increase, so that the STN structure is obtained without the splay structure. STN has an increased twist angle from 180° up to 270° [15–17], which is known to increase the steepness of the transmittance–voltage curve (V-T curve). In Figure 1, the V-T curve of LPSTN is shown as blue open triangles [9]. When the p/d ratio is 2.14, the RTN structure appears by application of a high voltage (metastable state 2 (ii) in Figure 1d), while the typical STN (p/d ratio is 1.5) does not change to RTN. Our group has termed this non-typical STN (p/d ratio more than 1.5) LPSTN.

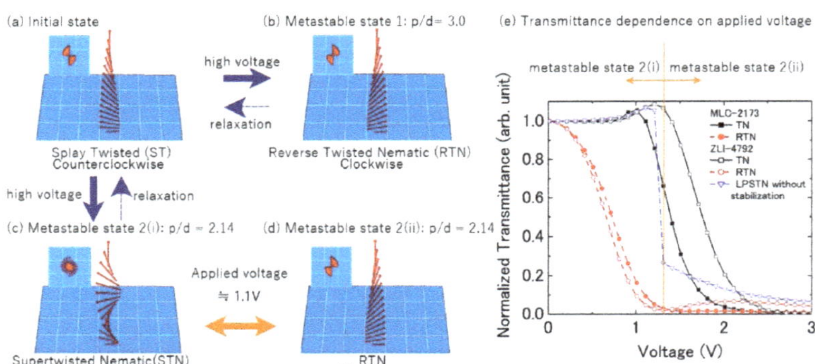

Figure 1. Schematic diagrams of LC structures; (**a**) initial state, ST; (**b**) metastable state 1, RTN; (**c**) metastable state 2a, LPSTN; (**d**) metastable state 2b, RTN. The ST structure changed to the metastable state (**b**) RTN or (**c**) LPSTN by high voltage application when the p/d ratio was 3.0 or 2.14. (**e**) Dependence of transmittance on the applied voltage for TN-LCDs (cell thickness of 5.0 μm), RTN-LCDs (chiral pitch of 15.0 μm, cell thickness of 5.0 μm, RM), and LPSTN-LCD (chiral pitch of 15.0 μm, cell thickness of 7.0 μm) [9,14]. Solid symbols indicate MLC-2173 and open symbols indicate ZLI-4792.

In this study, we focus on the effect of thermal energy and attempt to cause transformation by heat instead of voltage to maintain the structure by application of a low voltage because, by reducing the applied voltage, RTN-LCDs of high transmittance without applied voltage can be obtained. Transformation from the ST state to the RTN state occurred at a temperature slightly below the nematic–isotropic (NI) point with the application of 1.0 V. Furthermore, the electro–optical properties and Gibbs free energy were calculated for each structure and compared with the obtained results. The Gibbs free energy is calculated by

Frank–Oseen free energy density F_1 and the energy per unit volume by electric field F_2 F_1 is given by:

$$F_1 = \frac{1}{2}K_{11}(\nabla\cdot\hat{n})^2 + \frac{1}{2}K_{22}(\hat{n}\cdot\nabla\times\hat{n} + q_0)^2 + \frac{1}{2}K_{33}(\hat{n}\times\nabla\times\hat{n})^2 \quad (1)$$

where K_{11}, K_{22} and K_{33} are elastic constants for different director deformations: splay, twist and bend, respectively, $q_0 = 2\pi/P_0$ is constant and describes the chiral pitches P_0 of the helix, and $\hat{n}$ is eigenvector [18]. F_2 is given by:

$$F_2 = \frac{1}{2}\varepsilon E\cdot E \quad (2)$$

where ε is the dielectric tensor of the liquid crystal, and E is the electric field. The Gibbs free energy G can be expressed by:

$$G = F_1 - F_2 \quad (3)$$

2. Materials and Methods

The LC material MLC-2173 (US048) (Merck & Co. Inc., Darmstadt, Germany) had a left-handed pitch of 15.0 μm and MJ141067 (US048) had 5 wt% RM in MLC-2173 (US048). Two LC were mixed to be 1 wt% RM. The MLC-2173 has positive dielectric anisotropy (20.8). The polyimide material for the alignment layers was produced by mixing PIA-X768-01X and PIA-X359-01X (Chisso Petrochemical Corp., Tokyo, Japan) in a 15:85 ratio. The pretilt angle of the alignment layer was confirmed to be 6° with a pretilt analysis system (PAS-301, Elsicon Inc., Newark, NJ, USA). The cell thicknesses were set at 5.0 and 5.4 μm using silica bead spacers. The rubbing directions were set to form a right-handed twist of the LC material without a chiral reagent from the upper to the lower substrate. A splayed twisted structure was formed when left-handed LC materials were injected into the cell. The temperature was controlled with a high-precision temperature controller (mK1000, Instec Inc., Boulder, CO, USA) and UV light was irradiated to cure the RM with a UV lamp (LUV-6, AS ONE Co., Osaka, Japan). The electro-optical properties (applied voltage dependence of the transmittance and response time) were measured using OPTIPRO-standard (Shintech Inc., Yamaguchi, Japan). The optical axis of one polarizer was set parallel to one of the rubbing directions in a crossed nicols setting. The elastic constant was measured with an elastic constant measurement system (EC-1, Toyo Corp., Tokyo, Japan). Simulations of the electro-optical properties and Gibbs free energy were performed using LCD Master (Shintech Inc.) software under various conditions, such as pretilt angle, cell thickness, and chiral pitch. The LCDs were driven by a 1 kHz square wave.

3. Results and Discussions

The transition from ST to RTN was confirmed by polarized optical microscopy (POM) measurements under a crossed nicols condition. The clearing point for this LC is 353 K. Figure 2 shows the lowest sustain voltage for each cell thickness as a function of temperature. These voltages started to decrease near 343 K. Transition to the isotropic phase occurred with application of 1.0 V at 354.7 and 356.5 K for the cell thicknesses of 5.0 and 5.4 μm, respectively. The chiral pitch extended up to 16.8 μm at high temperature, as shown in Figure 3. The chiral pitch was once saturated near 343 K; however, it extended up to 16.8 μm at high temperature, as shown in Figure 3. The change of the molecular conformation of chiral reagent could occur at high temperature. The change of the pitch dependence on temperature could be explained by the change of the molecular conformation [19]. Therefore, the disturbance of the chiral pitch on temperature could be explained by the large change of the conformation of the chiral reagent. Although the reason is not stated, there are reports that the chiral pitch changed twice at high temperatures [20].

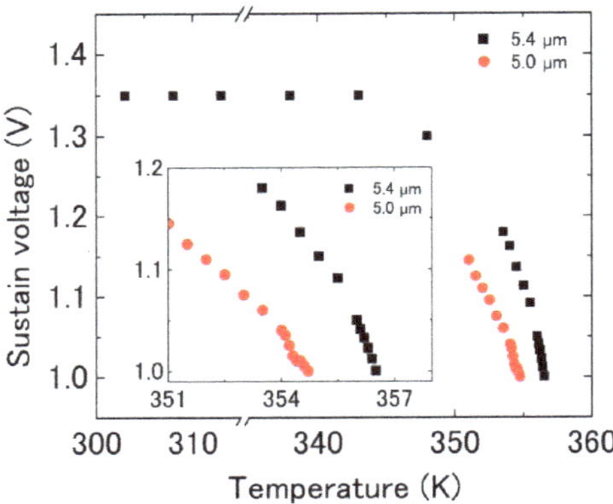

Figure 2. Sustain voltage for RTN structure as a function of temperature for cell thicknesses of 5.0 (red solid circles) and 5.4 μm (black solid squares). The insert shows an enlarged view of the high temperature region.

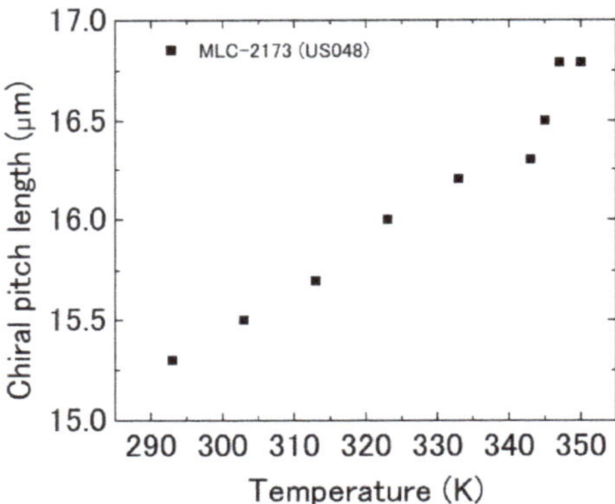

Figure 3. Chiral pitch length as a function of temperature.

The ST structure changed to the RTN structure with application of 1.2 V (transition voltage) at 356 K. A decrease in the voltage to 0.5 or 0 V resulted in a transition of the RTN structure to the ST structure via the STN state in Figure 4a,b. Transition speed to STN is faster than that to ST. At this time, the STN structure changed to the RTN structure with application of 1.0 V, while the ST structure was maintained. After a relatively long time of applying a voltage lower than 1.0 V, the structure becomes the ST structure (Figure 4c). It is necessary to apply 1.2 V to transform the ST structure into the RTN structure, while 1.0 V is required for a transition from the STN to RTN structure (Figure 4d). Therefore, there is a higher potential barrier between ST and RTN than that between STN and RTN.

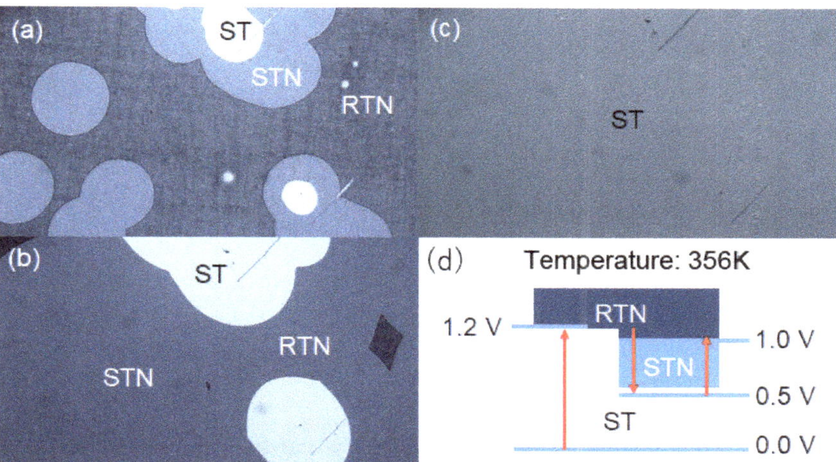

Figure 4. POM images acquired at 356 K; (**a**) immediately after changing the applied voltage from 1.2 to 0.5 V, (**b**) after a short time, and (**c**) after a relatively long time. (**a**) The structure changed from RTN to STN or ST. (**b**) The transition speed to STN was faster than that to ST. At this time, increasing the applied voltage to 1.0 V results in transition of the STN structure to the RTN structure, while the ST structure is maintained. (**c**) After a relatively long time of applying a voltage lower than 1.0 V, the structure eventually becomes the ST structure. (**d**) The ST structure changed to the RTN structure with application of 1.2 V at 356 K. A decrease in the voltage to 0.5 or 0 V resulted in a transition of the RTN structure to the ST structure via the STN state. The STN structure changed to the RTN structure with application of 1.0 V at 356 K.

These RTN-LCDs were irradiated with UV light for 1 h, while 1.0 V was applied at 356 K, after confirmation of the RTN structure with POM. After UV irradiation, the RTN structure was maintained at high temperature; however, the LPSTN structure was formed at room temperature when the cell thickness was 5.4 μm. Figure 5 (green solid triangles) shows V-T curves for the LPSTN structure. On the other hand, when the cell thickness was 5.0 μm, the RTN structure at high temperature relaxed to the ST structure at room temperature even after UV irradiation. The driving voltage increased in the order of RTN-LCD < LPSTN-LCD < TN-LCD. When the V-T curves of STN-LCD were calculated (LCD Master) for cell thicknesses of 5.0–6.2 μm with a chiral pitch of 15 μm, a pretilt of 6° and a twist angle of 270°, the driving voltage was high as a function of thickness (Figure 6). Figure 7 shows V-T curves calculated for various chiral pitches of 10–25 μm with a chiral pitch of 15 μm, a pretilt of 6°, a twist angle of 270°, and a cell thickness of 5.4 μm. A longer chiral pitch results in a lower driving voltage; however, if the pitch is too long (chiral pitch is 25 μm), then the transmittance slowly decreases as a function of the voltage. Although the transmittance of LPSTN did not decrease to zero at 1.2 V in Figure 5, the steepness of transmittance increased compared with the TN or RTN. In addition, the transmittance decreased to zero in simulation in the case of the cell thickness of 5.4 μm and chiral pitch of 20 μm. Therefore, we thought that a black state at low driving voltage can be realized by controlling condition of LCD.

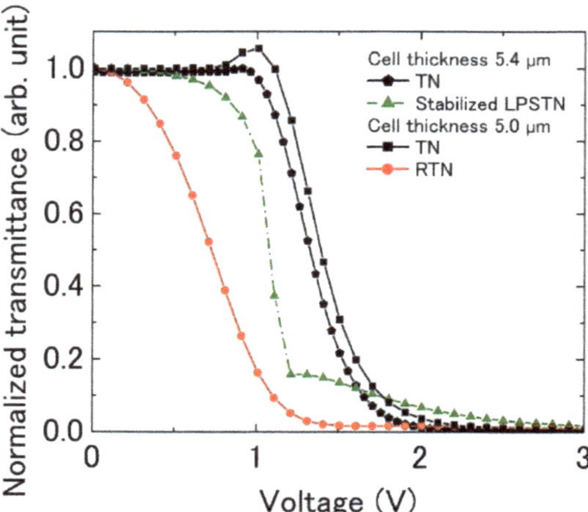

Figure 5. Transmittance as a function of applied voltage for TN (black solid pentagons, cell thickness of 5.4 μm), LPSTN (green solid triangles, chiral pitch of 15.0 μm, cell thickness of 5.0 μm, RM), TN (black solid squares, cell thickness of 5.0 μm), and RTN (red solid circles, chiral pitch of 15.0 μm, cell thickness of 5.0 μm, RM).

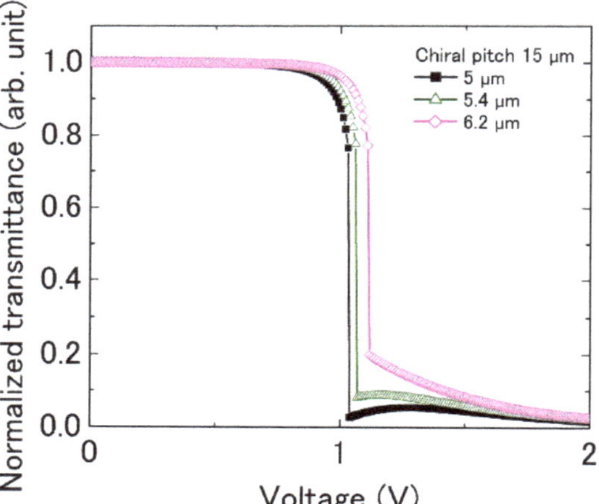

Figure 6. Transmittance as a function of the applied voltage for LPSTN (MLC-2173, chiral pitch 15.0 μm) with cell thicknesses of 5.0, 5.4, and 6.2 μm. The threshold and saturation voltage decreased with the cell thickness.

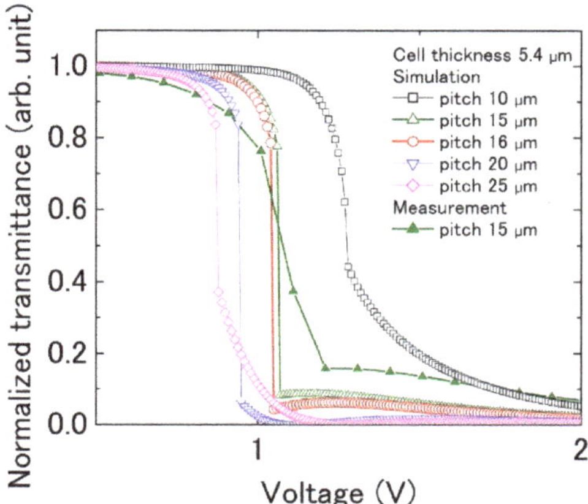

Figure 7. Transmittance as a function of applied voltage for LPSTN with chiral pitches of 5, 10, 15, 20, and 25 μm. Open symbols indicate simulation data and solid symbols show measurement data.

The transmittance and the change of azimuth angle dependence on voltage were calculated for LPSTN using LCD Master for a cell thickness of 5.4 μm and a chiral pitch of 15 μm, and the results are shown in Figures 8 and 9. The transmittance decreased similar to that for STN LCDs until 1.06 V and then dropped slowly from 1.07 V. The twist of the azimuth angle was 270° up to 1.06 V and then changed to 90°, which is not simple rotation. The LC molecules near the alignment layer twist left-handed where z/d implies the normalized thickness is from 0.0 to 0.4 and from 0.6 to 1.0 and those near the center of the cell twist right-handed at more than 90°. This phenomenon is an intrinsic twist for the RTN structure. Consequently, the LPSTN structure changed into an RTN structure with application over a certain voltage.

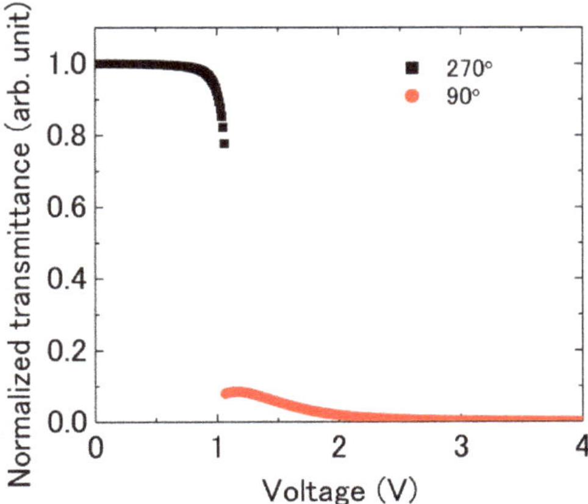

Figure 8. Transmittance as a function of the applied voltage for LPSTN (MLC-2173, chiral pitch 15.0 μm, cell thickness 5.4 μm).

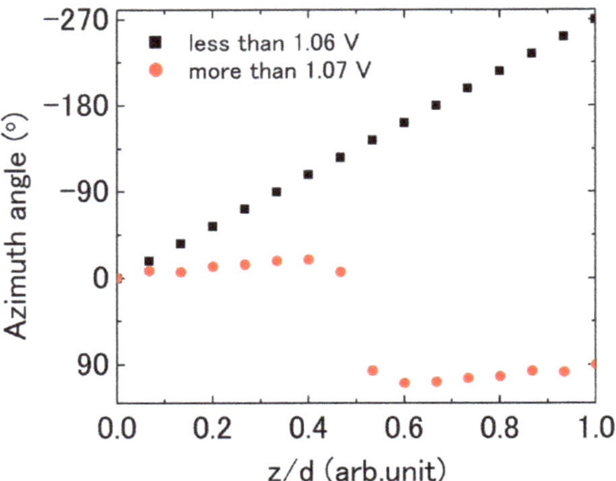

Figure 9. Azimuth angle dependence of the normalized cell thickness for LPSTN (MLC-2173, chiral pitch 15.0 μm, cell thickness 5.4 μm). The angle is 270° until the applied voltage is 1.06 V and 90° over 1.07 V. The z/d implies normalized thickness.

The chiral pitch is known to cause length increase, and the decrease in physical properties (e.g., the dielectric constant anisotropy, and the elastic constant) under heating. The chiral pitch was elongated to 16.8 μm (Figure 3) and the physical properties changed (Table 1) at 353 K. The physical properties at 353 K were estimated from multiplying nominal values at room temperature (298 K) by the ratio of measured values at 298 K to measured values at 353 K. The tilt angle in the vicinity of the alignment layer may be lower than that in the previous study [14], because the sustain voltage is low. The pretilt angle for LCDs has been reported to be controllable by polymer matrix formation in the vicinity of the alignment layers by application of an electric voltage [14,21,22].

Table 1. Dielectric constant anisotropies and elastic constants were measured at 298 and 355 K. The estimated value was calculated by dividing the nominal value by the ratio of measured values at 298 K to the measured values at 355 K.

	$\Delta\varepsilon$(F/m)	K_{11}(pN/m^2)	K_{22}(pN/m^2)	K_{33}(pN/m^2)
measured value@298 K	15.6	2.4	10.5	20.5
measured value@353 K	3.9	0.6	2.6	4.3
nominal value@298 K	20.8	8.7	18.1	13.3
estimated value@353 K	5.2	2.2	4.5	2.8

The pretilt angle was assumed 15° because the RTN structure was maintained by UV irradiation with application of 1 V, while the pretilt angle is 6.0° for MLC-2173 without chiral dopant and RM at room temperature. The Gibbs free energy for RTN and LPSTN was calculated for the difference of (i) the physical properties (Figure 10a,b), and (ii) the pretilt angles under the condition (i) (Figure 10c,d).

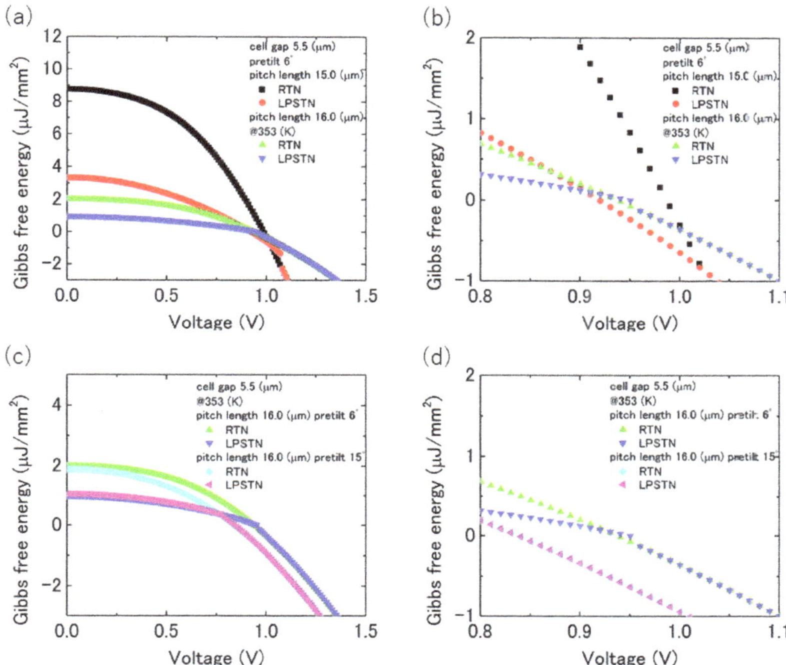

Figure 10. Gibbs free energy for the RTN and the LPSTN structures with a cell thickness of 5.5 μm (**a**) at 298 K or those at 353 K (**c**) at 353 K with a pretilt angle of 6° or 15° (lower left), and (**b**,**d**) enlarged views of the left-hand panels near 1 V.

At the beginning, the Gibbs free energy of the RTN and LPSTN structure at high temperature decreased at 0 V compared with the Gibbs free energy of those structures at room temperature (Figure 10a,b). Additionally, these energies then slowly decreased as a function of voltage. The transition voltage from LPSTN structure (blue inverted triangles) to RTN structure (green triangles) under the physical properties at 353 K was lower than the transition voltage from LPSTN structure (black squares) to RTN structure (red circles) at 298 K. Secondly, the transition voltage from LPSTN structure (sky blue diamonds) to RTN structure (purple left-pointing triangles) was lower than the transition voltage from LPSTN structure (blue inverted triangles) to RTN structure (green triangles) in Figure 10c,d. For application of 1.0 V, the RTN structure was stable because the LPSTN structure changed to the RTN structure, even if the pretilt angle was kept low.

For the RTN structure with the application of 1 V at 356 K experimentally, two routes are conceivable for the process of decreasing the voltage from 1 to 0 V; (i) the RTN remaining (green triangles in Figure 11) or (ii) the RTN changing to the LPSTN at 0.95 V (blue inverted triangles in Figure 11). The LPSTN structure was observed to change to the RTN structure with the application of 1.0 V at 356 K in Figure 4. Therefore, the RTN structure when polymerized with the application of 1 V at 356 K was the RTN structure that was changed from the LPSTN. After being polymerized, the LPSTN structure was stable at room temperature.

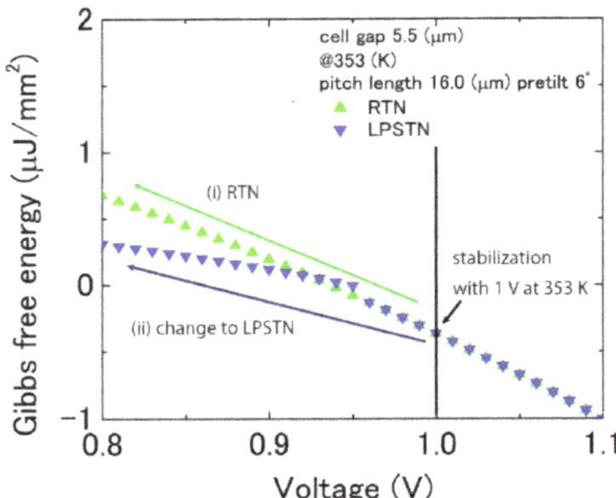

Figure 11. There are two routes from the RTN structure with application of 1 V at 353 K; (i) the RTN structure remaining, or (ii) changing to the LPSTN structure at low voltage.

A notable point of the STN-LCD is that their sharper threshold properties can realize a larger number of pixels in the case of passive matrix LCDs. For multiple-line addressing of passive matrix LCDs, the sharpness of the threshold property can be expressed by:

$$\alpha = \frac{V_{sat}}{V_{th}} \tag{4}$$

where V_{sat} is the saturation voltage and V_{th} is the threshold voltage. A sharper threshold property means that α approaches one more closely. The maximum number of scanning electrodes of passive matrix addressing N_{max}, can be expressed using α [23].

$$N_{max} = \left\{\frac{\alpha^2 + 1}{\alpha^2 - 1}\right\}^2 \tag{5}$$

In the case where α approaches one, the possible number of scanning electrodes increases. For this reason, high-resolution LCDs can be achieved using the sharp threshold property of the STN mode. However, because the STN mode uses the change in the retardation of liquid crystals, it has a problem in that the transmitted light is colored. An LPSTN LCD possesses the STN structure without an applied voltage and exhibits the sharp threshold property. However, when voltage larger than the threshold voltage is applied, it changes to the RTN-LCD structure and shows the black state without color. Therefore, an LPSTN can realize the sharp threshold property of the STN mode and the black state without the color of the TN mode.

The change from LPSTN to RTN occurs because the RTN and LPSTN states are considered topologically equivalent. The LPSTN mode showing low driving voltage and a realizing black state could be an improved alternative candidate for STN-LCDs.

4. Conclusions

An increase in the temperature decreased the voltage for the ST to RTN structural transition and the sustain voltage for the RTN structure. The LPSTN-LCD but not the RTN-LCD was successfully stabilized with UV irradiation near the clearing point. The differences between this study and previous studies are the temperature during UV irradiation and the cell thickness. The notable points of the LPSTN-LCD are a low driving voltage and a black display when a high voltage is applied under the crossed nicols condition. Due to the change from the STN structure to the TN structure, a black state can be realized, unlike the typical STN mode under the crossed nicols condition. The lower limit of the p/d ratio for the LPSTN was 2.14 in 2012; however, that was improved to 2.78 using this method. A lower driving voltage was calculated for chiral pitches of 15 or 20 μm and cell thicknesses of 5.0 or 5.4 μm (p/d ratios of 3.0 or 3.7). The structure was returned to the ST structure under p/d ratio of 3.0. If the p/d ratio is dominant, then the structure will be returned with p/d ratio of 3.7.

Author Contributions: M.I. proposed the devices and wrote the manuscript. K.T. supervised. S.O. prepared the sample cells and made measurements. All authors have read and agreed to the published version of the manuscript.

Funding: This research received no external funding.

Institutional Review Board Statement: Not applicable.

Informed Consent Statement: Not applicable.

Data Availability Statement: Not applicable.

Acknowledgments: The authors are thankful to Merck & Company Incorporated for providing the LC materials, and to Chisso Petrochemical Corporation for providing the polyimide materials. This work was supported by the Organization for Research Promotion Sanyo-Onoda City University.

Conflicts of Interest: The authors declare no conflict of interest.

References

1. Kim, Y.J.; Lee, S.-D. Reflective Mode of a Nematic Liquid Crystal with Chirality in a Hybrid Aligned Configuration. *Appl. Phys. Lett.* **1998**, *72*, 1978–1980. [CrossRef]
2. Kubono, A.; Kyokane, Y.; Akiyama, R.; Tanaka, K. Effects of Cell Parameters on the Properties of Hybrid Twisted Nematic Displays. *J. Appl. Phys.* **2001**, *90*, 5859–5865. [CrossRef]
3. Yamaguchi, R.; Sakamoto, Y. Homogeneous-Twisted Nematic Transition Mode Liquid Crystal Display by out-of-Plane Field. In Proceedings of the International Display Workshops (IDW'17), Sendai, Japan; 2017; Volume 24, pp. 294–295. Available online: https://researchmap.jp/read0188761/published_papers/19822221 (accessed on 9 December 2021).
4. Takatoh, K.; Kaneko, T.; Uesugi, T.; Kobayashi, S. New Approach to Reduce the Driving Voltage for TN-LCDs. In Proceedings of the International Display Workshop (IDW'07), Sappro, Japan; 2007; Volume 14, pp. 421–424. Available online: https://www.jstage.jst.go.jp/article/ekitou/2007/0/2007_0_103/_pdf/-char/ja (accessed on 9 December 2021).
5. Takatoh, K.; Akimoto, M.; Kaneko, H.; Kawashima, K.; Kobayashi, S. Molecular Arrangement for Twisted Nematic Liquid Crystal Displays Having Liquid Crystalline Materials with Opposite Chiral Structures (Reverse Twisted Nematic Liquid Crystal Displays). *J. Appl. Phys.* **2009**, *106*, 064514. [CrossRef]
6. Takatoh, K.; Uno, H.; Taniguchi, H.; Watanabe, I. Stable Reverse TN-LCDs (RTN) Using High Pretilt Angle Alignment Layers. In Proceedings of the International Display Workshops (IDW'16), Fukuoka, Japan; 2016; Volume 23, pp. 3–7. Available online: https://jglobal.jst.go.jp/en/detail?JGLOBAL_ID=202102285281164373 (accessed on 9 December 2021).
7. Takatoh, K.; Watanabe, I.; Yonezawa, Y.; Goda, K. The Novel TN LCD of the Lowest Driving Voltage. In Proceedings of the International Display Workshops (IDW'18), Nagoya, Japan; 2018; Volume 25, p. 211. Available online: https://www.idw.or.jp/IDW18AP.pdf (accessed on 9 December 2021).
8. Yamaguchi, R.; Kawata, S. Ultra-Low Driving Voltage in Quasi-Twisted Nematic Mode Using Weak/Strong Anchoring Hybrid Alignment Surface. In Proceedings of the International Display Workshops (IDW'20), Virtual, Japan; 2020; Volume 27, pp. 131–132. Available online: https://confit.atlas.jp/guide/event-img/idw2020/LCTp2-04/public/pdf_archive?type=in (accessed on 9 December 2021). [CrossRef]
9. Akimoto, M.; Minami, S.; San-No-Miya, M.; Kobayashi, S.; Takatoh, K. Different p/d Ratios Lead to Different Twisted LC Configurations in Reverse Twisted Nematic LCD. In Proceedings of the International Display Workshops (IDW'11), Nagoya, Japan; 2011; Volume 3, pp. 1593–1596. Available online: https://www.researchgate.net/publication/289406160_Different_pd_ratios_lead_to_different_twisted_LC_configurations_in_reverse_twisted_nematic_LCD (accessed on 9 December 2021).
10. John, V.N.; Shiju, E.; Arun, R.; Ravi Varma, M.K.; Chandrasekharan, K.; Sandhyarani, N.; Varghese, S. Effect of Ferroelectric Nanoparticles in the Alignment Layer of Twisted Nematic Liquid Crystal Display. *Opt. Mater.* **2017**, *67*, 7–13. [CrossRef]
11. SJ, S.; Gupta, R.K.; Kumar, S.; Manjuladevi, V. Enhanced Electro-Optical Response of Nematic Liquid Crystal Doped with Functionalised Silver Nanoparticles in Twisted Nematic Configuration. *Liq. Cryst.* **2020**, *47*, 1678–1690. [CrossRef]
12. Takatori, K.-I.; Sumiyoshi, K. Splayed TN Configuration Stability in Domain-Divided TN Mode. *Mol. Cryst. Liq. Cryst. Sci. Technol. Sect. A Mol. Cryst. Liq. Cryst.* **1995**, *263*, 445–458. [CrossRef]
13. Saito, Y.; Takano, H.; Chen, C.J.; Lien, A. Stability of UV-Type Two-Domain Wide-Viewing-Angle Liquid Crystal Display. *Jpn. J. Appl. Phys.* **1997**, *36*, 7216. [CrossRef]
14. Takatoh, K.; Watanabe, I.; Goda, K. Stabilisation of Reverse Twisted Nematic Structure by Forming Polymer Matrix in the Vicinity of the Alignment Layer. *Liq. Cryst.* **2020**, *47*, 1–5. [CrossRef]
15. Scheffer, T.J.; Nehring, J. A New, Highly Multiplexable Liquid Crystal Display. *Appl. Phys. Lett.* **1984**, *45*, 1021–1023. [CrossRef]

16. Li, J.; Lee, E.S.; Vithana, H.; Bos, P.J. Super-Twisted-Nematic Liquid Crystal Displays with Multi-Domain Structures. *Jpn. J. Appl. Phys.* **1996**, *35*, L1446. [CrossRef]
17. Scheffer, T.; Nehring, J. Supertwisted Nematic (Stn) Liquid Crystal Displays. *Annu. Rev. Mater. Sci.* **1997**, *27*, 555–583. [CrossRef]
18. Frank, F.C.I. Liquid Crystals. On the Theory of Liquid Crystals. *Discuss. Faraday Soc.* **1958**, *25*, 19–28. [CrossRef]
19. Huff, B.P.; Krich, J.J.; Collings, P.J. Helix Inversion in the Chiral Nematic and Isotropic Phases of a Liquid Crystal. *Phys. Rev. E* **2000**, *61*, 5372–5378. [CrossRef] [PubMed]
20. Shim, K.S.; Heo, J.U.; Jo, S.I.; Lee, Y.-J.; Kim, H.-R.; Kim, J.-H.; Yu, C.-J. Temperature-Independent Pitch Invariance in Cholesteric Liquid Crystal. *Opt. Express OE* **2014**, *22*, 15467–15472. [CrossRef] [PubMed]
21. Lu, L.; Sergan, V.; Bos, P.J. Mechanism of Electric-Field-Induced Segregation of Additives in a Liquid-Crystal Host. *Phys. Rev. E Stat. Nonlin Soft Matter Phys.* **2012**, *86*, 051706. [CrossRef] [PubMed]
22. Sergan, V.; Sergan, T.A.; Bos, P.J.; Lu, L.; Herrera, R.; Sergan, E.V. Control of Liquid Crystal Alignment Using Surface-Localized Low-Density Polymer Networks and Its Applications to Electro-Optical Devices. *J. Mol. Liq.* **2018**, *267*, 131–137. [CrossRef]
23. Lueder, E. *Liquid Crystal Displays: Addressing Schemes and Electro-Optical Effects*; Wiley: Hoboken, NJ, USA, 2001.

Article

Fabrication of Polarization Grating on *N*-Benzylideneaniline Polymer Liquid Crystal and Control of Diffraction Beam

Mizuho Kondo [1,2,*], Kyohei Fujita [1], Tomoyuki Sasaki [2,3], Moritsugu Sakamoto [2,3], Hiroshi Ono [2,3] and Nobuhiro Kawatsuki [1,2,*]

1 Department of Applied Chemistry, Graduate School of Engineering, University of Hyogo, 2167 Shosha, Himeji 671-2280, Japan; m1fujitak@gmail.com
2 Japan Science and Technology Agency (JST), Core Research for Evolutional Science and Technology (CREST), Chiyoda-ku, Tokyo 102-0076, Japan; sasaki_tomoy@vos.nagaoka.ac.jp (T.S.); sakamoto@vos.nagaokaut.ac.jp (M.S.); onoh@vos.nagaokaut.ac.jp (H.O.)
3 Department of Electrical Engineering, Nagaoka University of Technology, 1603-1 Kamitomioka, Nagaoka 940-2188, Japan
* Correspondence: mizuho-k@eng.u-hyogo.ac.jp (M.K.); kawatuki@eng.u-hyogo.ac.jp (N.K.); Tel.: +81-79-267-4014 (M.K.)

Abstract: Photoresponsive photoalignable liquid crystalline polymers composed of phenyl benzoate terminated with *N*-benzylideneaniline were evaluated. These polymers are capable of axis-selective photoreaction, photoinduced orientation, and surface relief grating formation. Polarization holography using an He-Cd laser beam at a wavelength of 325 nm demonstrated the formation of a surface relief grating with a molecularly oriented structure based on periodic light-induced reorientation and molecular motion. Electrical switching of diffracted light using an electric field response of twisted-nematic cell containing a low-molecular-weight liquid crystal in combination was also demonstrated.

Keywords: liquid crystalline polymer; axis-selective photoreaction; holography; beam diffraction

Citation: Kondo, M.; Fujita, K.; Sasaki, T.; Sakamoto, M.; Ono, H.; Kawatsuki, N. Fabrication of Polarization Grating on *N*-Benzylideneaniline Polymer Liquid Crystal and Control of Diffraction Beam. *Crystals* **2022**, *12*, 273. https://doi.org/10.3390/cryst12020273

Academic Editors: Jun Xu, Kohki Takatoh and Akihiko Mochizuki

Received: 21 January 2022
Accepted: 15 February 2022
Published: 17 February 2022

Publisher's Note: MDPI stays neutral with regard to jurisdictional claims in published maps and institutional affiliations.

1. Introduction

Surface alignment patterning of optically anisotropic materials has attracted substantial attention from the perspective of science and technology because of its potential application in photonic devices for polarized light communication. In particular, polarizing gratings (PGs) have been researched and developed widely in the field of polarized beam manipulation because of their high diffraction efficiency and polarization selectivity [1,2]. These are likely to be applied to beam steering [3–6], imaging [7,8], and near-eye displays [9,10], as well as 3D displays and virtual reality/augmented reality (VR/AR) [4,11–14]. A PG is a diffractive optical element in which the in-plane anisotropic orientation varies linearly along the surface and the magnitude of anisotropy is constant [15–23]. It is fabricated by methods such as optical orientation by two-flux interference exposure [24], micro-rubbing [25,26], photomasking [27], and direct beam writing [8,28]. In these methods, the diffraction efficiency is improved by increasing the variation in refractive index in the film or the non-uniform structure of the film. Liquid crystal polymers are suitable as materials for PGs because their birefringence can be modulated substantially by external stimuli. On this, various structures have been evaluated.

In the previous study, we studied the formation of PGs by two-flux interference exposure using azobenzene [29,30] and *N*-benzylideneaniline (NBA) [31,32]. PGs are formed under interference exposure owing to the periodic distribution of light intensity and polarization state. When exposed to a polarized light beam, the azobenzene and NBA portions parallel to the polarization direction of the light beam undergo photoisomerization, followed by a trans-cis-trans reorientation process to form an oriented structure. At this time, small optical anisotropy is generated in the liquid crystal polymer film. On the other hand, there is a significant increase in optical anisotropy when the photoirradiated film is heated above the glass transition temperature due to self-assemble mesogenic moiety with high mobility [33]. Furthermore, the NBA can be functionalized by a polymer reaction to deactivate the photoresponsive part or to increase the heat resistance by cross-linking [34]. In addition, NBA is more transparent to the visible light region than azobenzene, whereby it is more suitable for use in optical devices.

Recently, we reported an NBA polymer with a high orientation, birefringence, and high heat resistance [35]. This polymer has the capability to replace components with higher birefringence while maintaining orientation [36]. However, its application as a PG material has not been investigated. Herein, we prepared PG using this material and evaluated its diffraction properties.

2. Materials and Methods

2.1. Material

Polymethacrylate P1 coupled with phenyl benzoate terminated with NBA via a hexamethylene spacer was used (Figure 1a). P1 was synthesized by the method described in the previous literature [35]. The polymer exhibited a glass transition temperature (Tg) of 100 °C and retained its liquid crystalline phase above 295 °C. The films of the polymer were prepared by spin-coating a dichloromethane solution of the polymers onto quartz substrates.

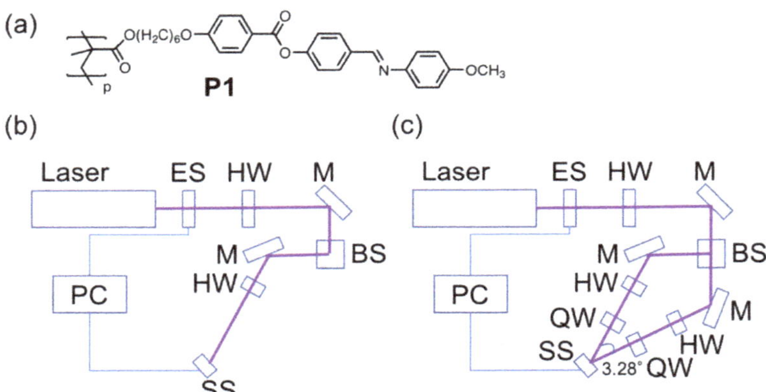

Figure 1. (**a**) Chemical structure of NBA liquid crystalline polymer P1. Experimental setup for (**b**) axis-selective photoreaction and (**c**) holographic exposure. ES, shutter; HW, half-wave plate; M, mirror; BS, beam splitter; QW, quarter wave plate; SS, sample stage; PC, personal computer.

2.2. Equipment

The photoinduced orientation behavior was evaluated by irradiating a 325 nm He-Cd laser (Kimmon IK3501R-G-S) (Tokyo, Japan) with the optical setup displayed in Figure 1b. Since the linearly polarized UV laser was emitted from the He-Cd laser source, the polarization direction of the actinic beam was controlled by a half-wave plate (HW) mounted on the rotational stage. The intensity of the laser beam was 210 mW/cm^2. The molecular orientation after linearly polarized light irradiation was evaluated using polarized absorption spectra (HITACHI U3900H) (Tokyo, Japan) and polarized light microscopy (Olympus BH-51) (Tokyo, Japan). The thermally annealed film was obtained by heating the photoirradiated film in the air at a specified temperature for 10 min.

Polarization holographic recording was performed using the experimental setup illustrated in Figure 1c. The laser beam was divided into two writing beams by a beam splitter (BS). The two writing beams with equal intensities crossed at an angle of 3.28° and impinged on the P1 films on a sample stage (SS). The resulting grating period of the polarization holographic grating was estimated to be approximately 4.8 μm. The polarization states of the two writing beams were controlled individually by two quarter-wave plates. The polarization holographic gratings were written using two orthogonally circularly polarized, mutually coherent He-Cd laser beams. The intensity of the interference beam was 420 mW/cm^2. The He-Ne laser with a wavelength of 633 nm was used as a probe light to evaluate the diffraction behavior, and the diffraction efficiency was calculated using the following equation:

$$DE = I_d / I_I$$

where I_d and I_I are the intensities of the diffracted and incident beams, respectively.

The laser intensity measurement, irradiation position, and dose control were performed using a serial communication program. The surface topography was measured using an interferometer-type surface profiler (Ryoka Systems, R3300H) (Osaka, Japan). A TN cell coated with indium tin oxide (ITO) and polyimide was used for the beam switching. A liquid crystal with a low molecular weight (Merck, ZLI4792: ($\Delta n = 0.09$, $\Delta\varepsilon = 5.3$) [37]) (Darmstadt, Germany) was injected into a TN cell (EHC KSRT-04/B211PINSS05) (Tokyo, Japan) with a cell gap of 4 μm. The orientation was controlled by applying a DC voltage of 2 V using a function generator (Hokuto Denko HB-111) (Tokyo, Japan).

3. Results and Discussion

3.1. Photoalignment Behavior in Thin Films

Figure 2 shows the variation in the polarized absorption spectra of the P1 film with a thickness of 140 nm when irradiated with a single linearly polarized laser beam for 10 s as shown in Figure 1b. The absorption at 340 nm originating from the π–π^* transition of NBA decreased in the direction parallel to the polarization of the laser and increased in the perpendicular direction. This indicated that the NBA moiety was rearranged perpendicular to the electric field vector of the polarized light by trans-cis-trans reorientation.

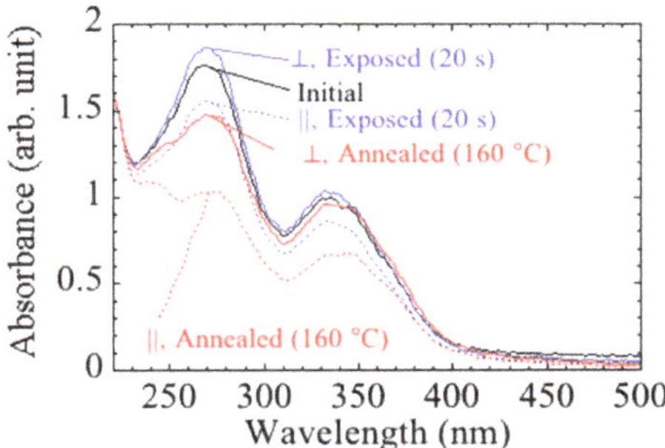

Figure 2. Polarized absorption spectra of P1 after spin-coating, irradiation with linearly polarized UV-laser, and annealing at 160 °C for 10 min. The spectra are color-coded as black for initial, blue for irradiation, and red for annealing, respectively. The absorption spectra for the directions parallel and perpendicular to the polarization direction of the laser beam are indicated by solid and dashed lines, respectively.

When heated to a temperature above the glass transition temperature, the optical anisotropy was enhanced perpendicular to the polarization direction of the laser beam. The amplified anisotropy was almost the same when the film was heated above 160 °C, while it showed a small value when the film was heated below 160 °C. The anisotropy induced by heat treatment tended to decrease for films exposed for more than 10 s. This may be due to undesired photoreactions (crosslinking or rearrangement reactions) and self-organization (out-of-plane orientation or aggregation) caused by prolonged exposure. The absorption spectra after annealing are plotted in red in Figure 2. These results are similar to those obtained under exposure to linearly polarized light from a mercury lamp [35].

Diffraction gratings were fabricated using interference exposure. Figure 3 plots the change in diffraction efficiency as a function of exposure time (solid line) and the change in diffraction efficiency of P1 films annealed at various temperatures after exposure to interference light (dashed line). The diffraction efficiency was dependent on the exposure time, increasing for exposure times shorter than 5 s and slowly decreasing for longer exposure times. The diffraction efficiency of the P1 film was stable after switching off the interfering beams though the diffraction efficiency after irradiation was very low (<0.5%). On the other hand, it was remarkably improved by heating above the glass transition temperature. The diffraction efficiency increased to 3.8% for the film exposed for 5 s at 240 °C. These can be attributed to the amplification of orientation and the expansion of birefringence modulation by heating and polarized laser irradiation. However, the diffraction efficiency was marginal (<5%), possibly due to the small film thickness.

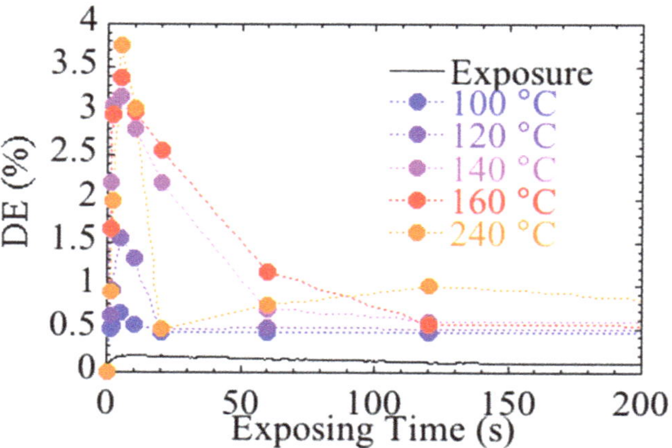

Figure 3. Variation in +1st diffraction efficiency of P1 film as a function of exposure time (solid line) and 1st diffraction efficiency of P1 film after thermal annealing under various temperatures (dashed lines).

The surface morphology of the heat-treated films was also evaluated. Periodic irregularities with a pitch of approximately 4 μm and a height difference of approximately 40 nm were observed after heat treatment. This indicated that the height difference was caused by the diffraction grating formation (Figure 4a). The surface relief was formed by the mass transfer of NBA upon light irradiation. This mechanism is identical to that of azobenzene and indicates that the diffractive structure is formed by a behavior identical to that of conventional NBA polymers. Figure 4b shows a polarized light micrograph of the diffraction grating. Periodic bright stripes are observed along the diffraction grating vector. This indicates that the NBA is arranged periodically and that the refractive index varies periodically. To investigate the optical anisotropy caused by the molecular arrangement, the grating was observed by rotating the polarizer and analyzer simultaneously by 30° while maintaining the grating fixed. The results showed that the rotation of the polarizer and analyzer caused the bright stripes to shift along the grating vector and recovered to the initial pattern at 180°. This indicates that the polarization direction of the linearly polarized light is modulated periodically by the interference of two orthogonal circularly polarized beams (Figure 4c) [38].

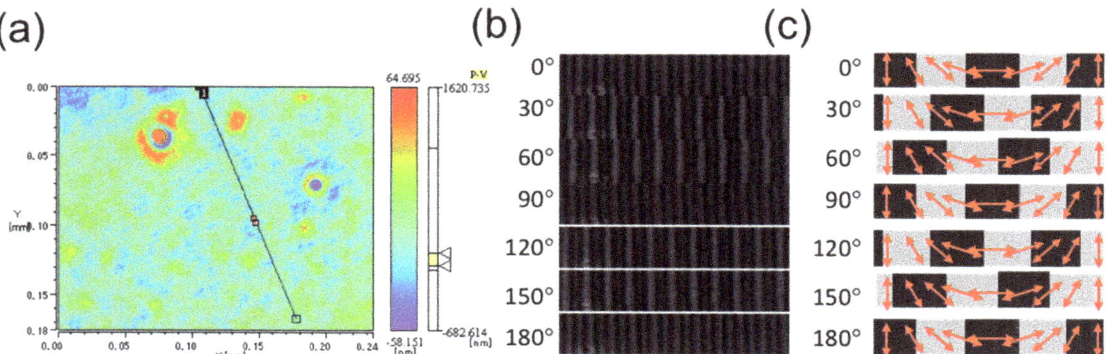

Figure 4. (**a**) Surface profile of P1 after irradiation with interference beams following thermal annealing (5 s, 240 °C). (**b**) Polarizing micrographs observed as a function of the rotation angle of the polarizer and analyzer, θ. (**c**) Alignment of mesogens induced in holograms. The arrows indicate the direction of mesogens. The white and gray areas correspond to the bright and dark areas, respectively, in (**c**).

3.2. Photoalignment Behavior in Thick Films

Figure 5a shows the polarization absorption spectra of a 460 nm-thick film after irradiation with LP-325 nm and heat treatment. Similar to the 140 nm film, the absorption in the direction parallel to the polarization field decreased with UV irradiation. However, a marginal amplification was induced by thermal annealing. Figure 5b shows the diffraction efficiencies of the films after interference exposure for various times. The diffraction efficiency after irradiation with the interference beam increased compared with that of the thin film. However, the increase after thermal treatment was marginal. Figure 5c shows the polarized light micrographs of the diffraction gratings. These show periodic bright stripes along the grating vectors as in the 140 nm film. However, the periodic structure is disrupted marginally after thermal treatment. The surface morphology of the film shows that the surface layer of the film became rougher after heat treatment. When this experiment was conducted using a significantly thicker film (900 nm), diffraction structures disappeared after heat treatment (data not shown). These results indicate that the thicker film increases the number of areas in the deeper layers of the film that are not involved in optical orientation. This, in turn, induces unintended self-organization and orientation (e.g., out-of-plane orientation), thereby resulting in surface roughening and a decrease in anisotropy. Meanwhile, reorientation is partially induced in the surface layer of the film by heat treatment. This is expected to increase the diffraction efficiency. It is necessary to reexamine the composition of the film (e.g., the transmittance and the introduction of a structure that facilitates reorientation) to increase the diffraction efficiency.

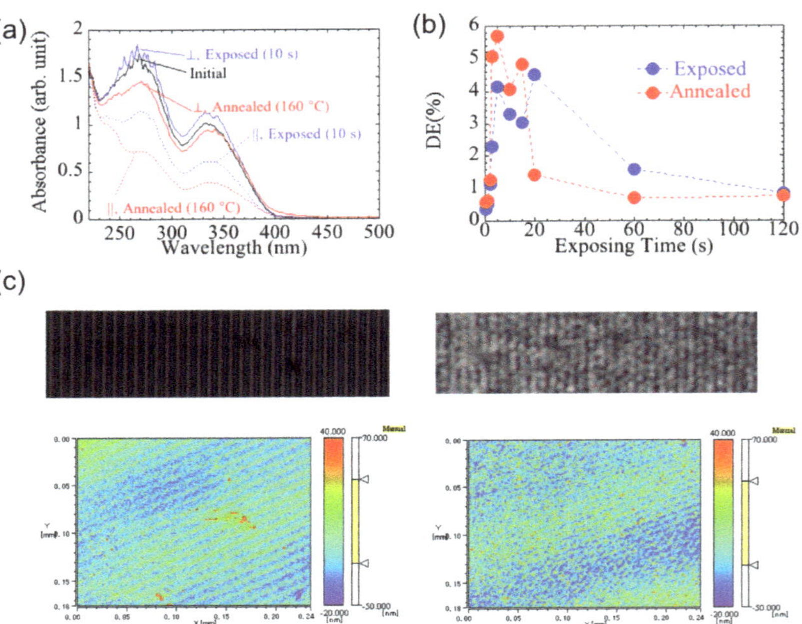

Figure 5. (**a**) Polarized absorption spectra of P1 film with a thickness of 460 nm after spin-coating, irradiation with linearly polarized UV-laser, and annealing at 160 °C for 10 min. The spectra are color-coded as black for initial, blue for irradiation, and red for annealing, respectively. The absorption spectra for the directions parallel and perpendicular to the polarization direction of the laser beam are indicated as solid and dashed lines, respectively. (**b**) Variation in +1st diffraction efficiency of P1 film with a thickness of 460 nm as a function of exposure time (solid line) and 1st diffraction efficiency of P1 film after thermal annealing at 160 °C. (**c**) POM image (**top**) and surface profile (**bottom**) of P1 film with thickness of 460 nm after irradiation (**left**) and thermal annealing (**right**).

3.3. Diffraction Control Using TN Cell

The combination of a field-driven cell and quarter-wave plate enables switching of the laser [39,40]. In particular, an optical system combining a field-driven TN cell and quarter-wave plate can switch the diffracted light from the +1st order to the −1st order, or vice versa, by controlling the voltage applied to the TN cell [4]. In this method, multiple diffraction angles can be obtained by cascading multiple PGs, and simultaneous wavelength control is feasible. Two-dimensional directional control is also likely to be feasible owing to the transmission type and cascade connection capability. Figure 6a,b shows the diffraction images of the voltage-off and voltage-on states, respectively (support Video S1). For the polarization grating, we used a 140 nm P1 film after thermal annealing. As expected, the diffraction order varies from +1 to −1 and vice versa. Since it is intended to be controlled by a single-board computer, DC voltage control is used; however, simple AC voltage (4 V, 60 Hz) can also be used for control (support Figure S1). Efficient switching is likely to be achieved by improving the diffraction efficiency. This is presently under study.

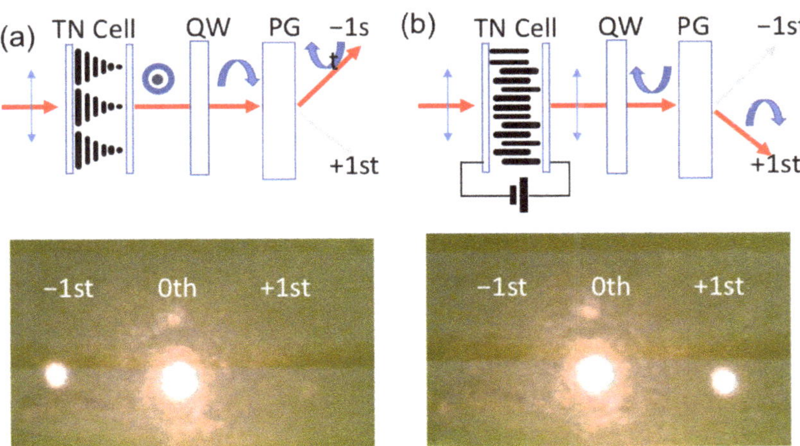

Figure 6. Schematic diagram (**top**) and photograph of diffraction (**bottom**) of proposed optical switch based on PG at (**a**) voltage-off state and (**b**) voltage-on state.

4. Conclusions

Axis-selective photoreaction, photoinduced orientation, and surface relief grating formation were evaluated using liquid crystalline polymers composed of phenyl benzoate terminated with NBA. Heat treatment greatly amplifies the molecular arrangement in the photo-reactive thin film, while the amplification is small in the photo-reactive thick film. The formation of a polarization grating with a molecularly oriented structure based on periodic light-induced reorientation and molecular motion was demonstrated. We demonstrated the electrical switching of diffracted light using a TN cell. Although the transmission of zero-order light could not be prevented in this study, more efficient beam control is likely to be achieved by biasing the diffraction of first-order light through the formation of a more efficient diffraction grating. It is necessary to reexamine the composition of the film, such as the transmittance and the introduction of a structure that facilitates reorientation.

Supplementary Materials: The following are available online at https://www.mdpi.com/article/10.3390/cryst12020273/s1, Figure S1: Diffraction photographs of an optical switch using an AC input (4 V, 60 Hz), Video S1: Switching of diffracted beam using field-driven cell.

Author Contributions: Conceptualization, N.K. and M.K.; Methodology, M.K.; Software, M.K.; Validation, K.F., M.K.; Investigation, K.F.; Optical setup, H.O., T.S., M.S.; Resources, N.K.; Writing—original draft preparation, M.K.; Writing—review and editing, N.K.; Visualization, M.K.; Supervision, N.K. All authors have read and agreed to the published version of the manuscript.

Funding: This research received no external funding.

Acknowledgments: This work was supported by Japan Science and Technology Agency (CREST JPMJCR2101).

Conflicts of Interest: The authors declare no conflict of interest.

References

1. Lin, T.; Xie, J.; Zhou, Y.; Zhou, Y.; Yuan, Y.; Fan, F.; Wen, S. Recent Advances in Photoalignment Liquid Crystal Polarization Gratings and Their Applications. *Crystals* **2021**, *11*, 900. [CrossRef]
2. Nersisyan, S.R.; Tabiryan, N.V.; Steeves, D.M.; Kimball, B. Optical Axis Gratings in Liquid Crystals and their use for Polarization insensitive optical switching. *J. Nonlinear Opt. Phys. Mater.* **2009**, *18*, 1–47. [CrossRef]
3. Oh, C.; Kim, J.; Muth, J.; Serati, S.; Escuti, M.J. High-Throughput Continuous Beam Steering Using Rotating Polarization Gratings. *IEEE Photonics Technol. Lett.* **2010**, *22*, 200–202. [CrossRef]

4. Chen, H.; Weng, Y.; Xu, D.; Tabiryan, N.V.; Wu, S.-T. Beam steering for virtual/augmented reality displays with a cycloidal diffractive waveplate. *Opt. Express* **2016**, *24*, 7287–7298. [CrossRef]
5. Tabirian, N.V.; Roberts, D.; Liao, Z.; Ouskova, E.; Sigley, J.; Tabirian, A.; Slagle, J.; McConney, M.; Bunning, T.J. Size, weight, and power breakthrough in nonmechanical beam and line-of-sight steering with geo-phase optics. *Appl. Opt.* **2021**, *60*, G154–G161. [CrossRef]
6. Kim, J.; Oh, C.; Serati, S.; Escuti, M.J. Wide-angle, nonmechanical beam steering with high throughput utilizing polarization gratings. *Appl. Opt.* **2011**, *50*, 2636–2639. [CrossRef]
7. Kudenov, M.W.; Escuti, M.J.; Dereniak, E.L.; Oka, K. White-light channeled imaging polarimeter using broadband polarization gratings. *Appl. Opt.* **2011**, *50*, 2283–2293. [CrossRef]
8. Ono, H.; Wada, T.; Kawatsuki, N. Polarization imaging screen using vector gratings fabricated by photocrosslinkable polymer liquid crystals. *Jpn. J. Appl. Phys.* **2012**, *51*, 030202. [CrossRef]
9. Xiong, J.; Tan, G.; Zhan, T.; Wu, S.-T. Wide-view augmented reality display with diffractive cholesteric liquid crystal lens array. *J. Soc. Inf. Disp.* **2020**, *28*, 450–456. [CrossRef]
10. Zhan, T.; Zou, J.; Xiong, J.; Liu, X.; Chen, H.; Yang, J.; Liu, S.; Dong, Y.; Wu, S.-T. Practical Chromatic Aberration Correction in Virtual Reality Displays Enabled by Cost-Effective Ultra-Broadband Liquid Crystal Polymer Lenses. *Adv. Opt. Mater.* **2020**, *8*, 1901360. [CrossRef]
11. Zhan, T.; Lee, Y.-H.; Wu, S.-T. High-resolution additive light field near-eye display by switchable Pancharatnam–Berry phase lenses. *Opt. Express* **2018**, *26*, 4863–4872. [CrossRef]
12. Zhan, T.; Xiong, J.; Tan, G.; Lee, Y.-H.; Yang, J.; Liu, S.; Wu, S.-T. Improving near-eye display resolution by polarization multiplexing. *Opt. Express* **2019**, *27*, 15327–15334. [CrossRef]
13. Xiong, J.; Wu, S. Rigorous coupled-wave analysis of liquid crystal polarization gratings. *Opt. Express* **2020**, *28*, 35960–35971. [CrossRef]
14. Wang, Q.-H.; Ji, C.-C.; Li, L.; Deng, H. Dual-view integral imaging 3D display by using orthogonal polarizer array and polarization switcher. *Opt. Express* **2016**, *24*, 9–16. [CrossRef]
15. Lee, Y.-H.; Yin, K.; Wu, S.-T. Reflective polarization volume gratings for high efficiency waveguide-coupling augmented reality displays. *Opt. Express* **2017**, *25*, 27008–27014. [CrossRef]
16. Packham, C.; Escuti, M.; Ginn, J.; Oh, C.; Quijano, I.; Boreman, G. Polarization Gratings: A Novel Polarimetric Component for Astronomical Instruments. *Public. Astron. Soc. Pac.* **2011**, *122*, 1471–1478. [CrossRef]
17. Nikolova, L.; Todorov, T. Diffraction efficiency and selectivity of polarization holographic recording. *Opt. Acta* **1984**, *31*, 579–588. [CrossRef]
18. Bomzon, Z.; Biener, G.; Kleiner, V.; Hasman, E. Space-variant Pancharatnam-Berry phase optical elements with computer-generated subwavelength gratings. *Opt. Lett.* **2002**, *27*, 1141–1143. [CrossRef]
19. Gori, F. Measuring Stokes parameters by means of a polarization grating. *Opt. Lett.* **1999**, *24*, 584–586. [CrossRef]
20. Roberts, D.; Kaim, S.; Tabiryan, N.V.; McConney, M.; Bunning, T.J. Polarization-independent diffractive waveplate optics. In Proceedings of the 2018 IEEE Aerospace Conference, Big Sky, MT, USA, 3–10 March 2018; pp. 1–11.
21. Tabiryan, N.; Roberts, D.; Steeves, D.; Kimball, B. 4G Optics: New Technology Extends Limits to the Extremes. *Photonics Spectra* **2017**, *51*, 46–50.
22. Tabiryan, N.V.; Roberts, D.E.; Liao, Z.; Hwang, J.; Moran, M.; Ouskova, O.; Pshenichnyi, A.; Sigley, J.; Tabirian, A.; Vergara, R.; et al. Advances in Transparent Planar Optics: Enabling Large Aperture, Ultrathin Lenses. *Adv. Opt. Mater.* **2021**, *9*, 2001692. [CrossRef]
23. Nersisyan, S.R.; Tabiryan, N.V.; Steeves, D.M.; Kimball, B.R. The Promise of Diffractive Waveplates. *Opt. Photonics News* **2010**, *21*, 41–45. [CrossRef]
24. Zhan, T.; Lee, Y.-H.; Tan, G.; Xiong, J.; Yin, K.; Gou, F.; Zou, J.; Zhang, N.; Zhao, D.; Yang, J.; et al. Pancharatnam-Berry optical elements for head-up and near-eye displays. *J. Opt. Soc. Am. B* **2019**, *36*, D52–D65. [CrossRef]
25. Honma, M.; Nose, T. Twisted nematic liquid crystal polarization grating with the handedness conservation of a circularly polarized state. *Opt. Express* **2012**, *20*, 18449–18458. [CrossRef] [PubMed]
26. Honma, M.; Nose, T. Highly efficient twisted nematic liquid crystal polarization gratings achieved by microrubbing. *Appl. Phys. Lett.* **2012**, *101*, 041107. [CrossRef]
27. Hisano, K.; Ota, M.; Aizawa, M.; Akamatsu, N.; Barrett, C.J. Shishido, A. Single-step creation of polarization gratings by scanning wave photopolymerization with unpolarized light. *J. Opt. Soc. Am. B* **2019**, *36*, D112–D118. [CrossRef]
28. Noda, K.; Kawai, K.; Sasaki, T.; Kawatsuki, N.; Ono, H. Multilevel anisotropic diffractive optical elements fabricated by means of stepping photo-alignment technique using photo-cross-linkable polymer liquid crystals. *Appl. Opt.* **2014**, *53*, 2556. [CrossRef]
29. Emoto, A.; Wada, T.; Shioda, T.; Sasaki, T.; Manabe, S.; Kawatsuki, N.; Ono, H. Vector gratings fabricated by polarized rotation exposure to hydrogen-bonded liquid crystalline polymers. *Jpn. J. Appl. Phys.* **2011**, *50*, 032502. [CrossRef]
30. Sasaki, T.; Izawa, M.; Noda, K.; Nishioka, E.; Kawatsuki, N.; Ono, H. Temporal formation of optical anisotropy and surface relief during polarization holographic recording in polymethylmethacrylate with azobenzene side groups. *Appl. Phys. B Lasers Opt.* **2014**, *114*, 373–380. [CrossRef]

31. Sasaki, T.; Nishioka, E.; Noda, K.; Kondo, M.; Kawatsuki, N.; Ono, H. Analysis of Non-Sinusoidal Surface Relief Structures Induced by Elliptical Polarization Holography on Azobenzene-Containing Polymeric Films. *Jpn. J. Appl. Phys.* **2014**, *53*, 02BB06. [CrossRef]
32. Kawatsuki, N.; Hosoda, R.; Kondo, M.; Sasaki, T.; Ono, H. Molecularly oriented surface relief formation in polymethacrylates comprising N-benzylideneaniline derivative side groups. *Jpn. J. Appl. Phys.* **2014**, *53*, 128002. [CrossRef]
33. Kawatsuki, N. Photoalignment and Photoinduced Molecular Reorientation of Photosensitive Materials. *Chem. Lett.* **2011**, *40*, 548–554. [CrossRef]
34. Ito, A.; Norisada, Y.; Inada, S.; Kondo, M.; Sasaki, T.; Sakamoto, M.; Ono, H.; Kawatsuki, N. Photoinduced Reorientation and Photofunctional Control of Liquid Crystalline Copolymers with in Situ-Formed N-Benzylideneaniline Derivative Side Groups. *Langmuir* **2021**, *37*, 1164–1172. [CrossRef] [PubMed]
35. Nisizono, T.; Kondo, M.; Kawatsuki, N. Photoinduced Molecular Reorientation of a Liquid Crystalline Polymer with a High Birefringence. *Chem. Lett.* **2021**, *50*, 912–915. [CrossRef]
36. Sakai, A.; Nnishizono, T.; Kondo, M.; Sasaki, T.; Sakamoto, M.; Ono, H.; Kawatsuki, N. Birefringence control of photoalignable liquid crystalline polymers based on an in situ exchange of oriented mesogenic side groups. *Chem. Lett.* **2021**, *50*, 91–93. [CrossRef]
37. Yamaguchi, R.; Sasaki, R.; Inoue, K. Driving Voltage in Reverse Mode Cell Using Reactive Mesogen: Effect of UV Absorption of Liquid Crystal. *J. Photopolm. Sci.* **2018**, *31*, 301. [CrossRef]
38. Shishido, A.; Ishiguro, M.; Ikeda, T. Circular Arrangement of Mesogens Induced in Bragg-type Polarization Holograms of Thick Azobenzene Copolymer Films with a Tolane Moiety. *Chem. Lett.* **2007**, *36*, 1146–1147. [CrossRef]
39. Sarkissian, H.; Serak, S.V.; Tabiryan, N.V.; Glebov, L.B.; Rotar, V.; Zeldovich, B.Y. Polarization- controlled switching between diffraction orders in transverse-periodically aligned nematic liquid crystals. *Opt. Lett.* **2006**, *31*, 2248–2250. [CrossRef]
40. Wu, S.T.; Efron, U.; Hess, L.D. Birefringence measurements of liquid crystals. *Appl. Opt.* **1984**, *23*, 3911–3915. [CrossRef]

MDPI

Article

Electrical Control of Optical Liquid-Crystal-Guided Microstructures

Michal Kwasny and Urszula A. Laudyn *

Faculty of Physics, Warsaw University of Technology, Koszykowa 75, 00-662 Warsaw, Poland; michal.kwasny@pw.edu.pl

* Correspondence: urszula.laudyn@pw.edu.pl; Tel.: +48-22-234-72-77

Abstract: This work investigates nematic liquid crystal (NLC) optical guiding structures designed in typical sandwich-like NLC cells. With the support of an electrically controlled spatial topology of director orientation, we manage a linear and nonlinear light propagation with the realization of optical beam switching.

Keywords: nematic liquid crystals; electro-optical beam steering; optical waveguides and microstructures

Citation: Kwasny, M.; Laudyn, U.A. Electrical Control of Optical Liquid-Crystal-Guided Microstructures. *Crystals* **2022**, *12*, 325. https://doi.org/10.3390/cryst12030325

Academic Editors: Kohki Takatoh, Jun Xu and Akihiko Mochizuki

Received: 1 February 2022
Accepted: 23 February 2022
Published: 26 February 2022

1. Introduction

In most optical waveguide designs with external modulators, the optical switching is commonly realized through electro-optic, magneto-optic, all-optical, and thermo-optical effects [1]. The classic electro-optic (EO) effect is related to the refractive index alteration due to the modification of the index ellipsoid (or optical indicatrix) by applying an external electric field. Commonly used materials for electro-optic waveguides are $LiNbO_3$, $LiTaO_3$, $BaTiO_3$, electro-optic polymers, and nematic liquid crystals (NLCs) [1–6]. Due to the very high birefringence, combined with the possibility of changing it under the effect of external stimuli in thin-film NLC elements, the latter offers an excellent opportunity to develop novel methods and devices for the control of light beams [6,7]. The fluid nature of NLCs and their compatibility with most optoelectronic materials, polymers, and organic materials allow them to be easily incorporated with other elements in various configurations, forms, and geometries, thereby increasing the potential applications in novel photonic networks. NLCs are composed of rod-like molecules. Due to anisotropy in molecular structures, they exhibit electrical anisotropy. In a specific temperature range, called the nematic phase, long axes of molecules are approximately parallel to each other in. The averaged alignment direction defines a dimensionless unit vector $\boldsymbol{n}$ called the director. Most nematics are optically uniaxial and positive birefringent materials (with an extraordinary refractive index greater than the ordinary one, $n_e > n_o$) with an optical axis corresponding to the long axis of the molecules. The electric field oscillations are perpendicular and parallel to the direction of molecular orientation, and then an ordinary and extraordinary wave can be excited, respectively.

A standard configuration of the planar NLC cell that comprises two glass substrates and a liquid crystal layer is used. Light beams are directed perpendicular to the substrates. The transmission of light beams directed perpendicular to the substrates of the cell can be electrically controlled. Several optical types of NLC devices were demonstrated, using the linear properties of the material and electrical switching. The simplest example is the pixelated microdisplay (i.e., a spatial light modulator) for free-space beam shaping and steering [8]. Different possibilities are associated with the light beams reflected from the spatially modulated NLC structures. Spatially structured NLC cells are also used to transform linearly polarized light beams into beams with radial or azimuthal polarization [9]. In the last decades, another geometry in which the light beam propagates in the NLC layer along the glass substrates has also attracted considerable attention [10,11]. The concept of

NLC waveguide channels for polarization-independent light propagation was presented, and, for the first time, the concept and performance of a liquid-crystal-based electro-optical router was discussed. In addition to electro-optic and nonlinear effects that can be utilized for controlling the light propagation in photonic channels, the optical switching between waveguides can be realized using ferroelectric or smectic liquid crystals [10,12]. The presented waveguides consisted of an NLC infiltrated core made in PDMS channels. Such a configuration uses NLC in their isotropic phase for electro-optical-induced waveguides and supports polarization-independent light propagation [11]. Such NLC-waveguide architectures, as reported by the Fraunhofer IPMS [13], are based on a pure electro-optic approach. On the other hand, it has to be underlined that liquid crystal devices can also switch and route signals due to the modulation of external electric fields and the interplay between the light and molecules [14], especially when the nonlinearity of NLC plays a crucial role [15]. It has been shown in several papers that a milliwatt of power of laser radiation is sufficient for observation of the self-focusing effects and formation of the soliton-type propagation of light beams due to reorientation of the NLC director under an electromagnetic field of an optical wave [16–19]. Such soliton beams, called nematicons, form a waveguide channel in NLC film that can confine and guide the light beam. Two basic configurations can be distinguished: (i) Electro-optical, which combines initial reorientation under the external electric field with a huge nonlinear optical reorientation of the liquid crystal molecules, as well as the electrically induced reorientation of molecules with a substantial nonlinear optical response of the liquid crystal molecules. It can be realized by applying a voltage over a liquid crystal layer between a planar electrode and a thin stripe electrode [17,18]. (ii) Pure all-optical, which combines a proper initial boundary condition with nonlinear optical reorientation [20]. It is possible to perform all-optical switching and logic gating using only interactions between nematicons [21]. The possibility of confining a weak signal beam in the self-induced waveguide [22] opens the way to build novel optical interconnects for computing and communications. A few results of soliton manipulation have been proposed to control the soliton trajectory, e.g., the temperature [23,24] or initial boundary conditions [25,26].

This work utilizes light beam propagation in NLC media and demonstrates an electro-optical architecture that provides voltage control over the optical path. The operation principle of the presented device is based on the electrically controlled spatial topology of director orientation due to the reorientation of NLC molecules in the presence of a low-frequency electric field and specially designed electrode geometry. Our concept of reconfigurable optical microstructures supports the waveguiding of Gaussian beams in both linear and nonlinear cases, the latter in the sense of nematicons. In a linear case, an optical waveguide is induced by an electric field across the cell, within regions delimited by stripe, arch, or Y-shape electrodes. Simple waveguides are easily combined in more complex structures, including independently addressed electrodes providing light beam guiding, switching, and steering. In the nonlinear regime, we successfully demonstrate the complete electro-optical control by combining the electric response of NLC molecules with nonlinear light beam propagation in the sense of nematicons and utilizing the NLC cell with properly designed individually addressed electrodes over nematicons' trajectory. Both approaches can potentially obtain light-guiding structures in a planar geometry.

2. Materials and Methods

The basic waveguide structure of the electrically controlled NLC is shown in Figure 1a. It consists of two glass substrates, numbered 1 and 2, glued together with a specific distance. On the first substrate, there is a transparent indium/tin oxide electrode. On the second one, metallic electrodes of different sizes and shapes are prepared by the electron beam lithography (EBL) technique. The preparation of the electrodes is a two-stage lift-off process. Initially, a 200 nm thin poly(methyl-methacrylate) layer (PMMA 495K) is spin-coated on the clean glass substrate. After evaporation of the solvent, an additional gold layer is deposited on the surface of the polymer. This step is essential because a thin layer of gold ensures the electrical discharging of the sample exposed on a focused beam of electrons, which irradiates the target shapes for the electrodes. Before the development process (in 1:3 solution of methyl isobutyl ketone:2-Propanol), a gold layer is removed by immersing the sample in KI_3 gold etchant (Sigma-Aldrich, St. Louis, MI, USA) and deionized water. The next stage of the process concerns subsequent metallization with a 10 nm underlay of Titanium and a 100 nm layer of gold. A final result is a glass substrate with deposited gold electrodes in designed shape and dimensions.

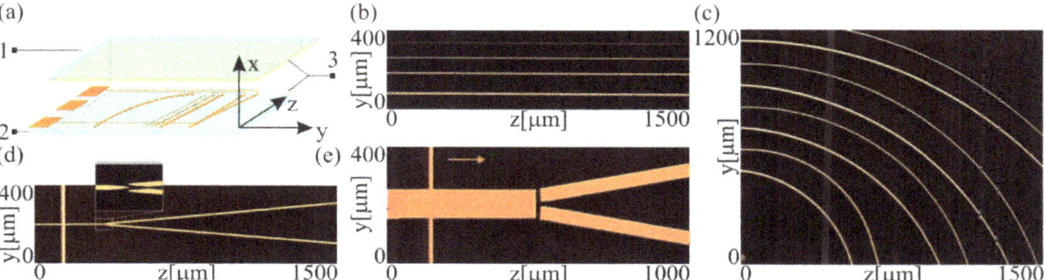

Figure 1. Electrically controlled NLC structures: (**a**) sketch of the NLC cell; (**b**) straight and (**c**) curved electrodes (widths 5 μm and 10 μm); (**d**) combination of three independently controlled electrodes for realizing optical Y-junction (10 μm width, one input and two output ports); (**e**) combination of Y-shaped independently controlled electrodes (the widths 100 μm and 50 μm) for the realization of beam switching in nonlinear propagation regime (a nematicon).

Whenever required, the additional polymeric layer (number 3, Figure 1a) can be spin-coated on the substrates on the top of the electrodes and their surrounded areas and rubbed to ensure the initial planar orientation in the desired direction. Such prepared glass substrates are then used for cell preparation. The cell thickness equals 12 and 30 μm for linear and nonlinear beam propagation, respectively. A small amount of NOA-61 UV-curing photopolymer mixed with 12 or 30 μm glass spacers is placed over one substrate. The counter substrate is put over it, and after proper alignment, the photopolymer is cured by exposing it to ultraviolet light. Because of the relatively small cell thickness used for linear beam propagation and electro-optical waveguide formation, no additional interfaces limiting the outflow of NLC were used, yet no adverse effects related to NLC leakage or meniscus formation were observed. The NLC cells that are made contain electrodes to realize the electrically induced waveguides for the light beam propagation and switching between two output ports. Prepared structures are shown in Figure 1b (straight electrodes), Figure 1c (bend electrodes), and Figure 1d ("fork" electrode), and they are designed to operate in a linear optical regime, i.e., when the intensity of the propagated beam is not high enough to cause additional reorientation of the molecules. The width of the electrodes is 5 and 10 μm; the length of the electrodes is 4 mm (straight); 0.5–1.5 mm (bend); 0.35 mm (the length of the input electrode of the fork structure); and 1.2 mm is the length after forking. The minimal distance between electrodes is 3 μm.

For nonlinear beam switching, an NLC cell with one electrode of a 100 μm width and 500 μm length and two output electrodes of a 50 μm width, 500 μm length and minimal separation of 10 μm is used, as shown in Figure 1e. This cell contains a rubbed alignment layer that anchors molecules of NLC along the z-axis.

The assembled cells are filled up with NLC by capillarity, taking care to avoid bubbles/gaps near the boundaries, and examined under a polarization microscope to verify the proper alignment of the director. As an NLC material, a typical-birefringent 6CHBT nematic liquid crystal is used: characterized by positive dielectric anisotropy ($\Delta\epsilon = 0.42$ at 1 kHz); refractive indices: $n_o = 1.4967$, $n_e = 1.6335$ at $\lambda = 1064$ nm and $n_o = 1.52$, $n_e = 1.68$ at $\lambda = 532$ nm (at room temperature); the Frank elastic constants are $K_{11} = 8.96$ pN;

K_{22} = 3.61 pN; K_{33} = 9.71 pN (for splay, twist and bend, respectively); and the nematic-isotropic phase transition temperature is T_c = 43 °C [27].

Figure 2a presents the basic experimental setup used for beam propagation analysis. As an input, we used the linearly polarized (along the *x*-axis) Gaussian beam (TEM_{00}) propagating along the *z*-axis. The beam is focused in the cell midplane (a beamwidth w_0 = 2.5 μm) at the input of the cell (z = 0). The first CCD camera visualized the beam evolution along z in the principal *yz*-plane by imaging the light scattered out of the plane, while the second CCD camera is used to visualize the beam profile in the *xy*-plane at the output. The position of the sample is adjusted by the precise three-axis micro-translation stage in the *x*, *y*, and *z*-direction. A low-frequency (1 kHz) sinusoidal waveform generator combined with a signal amplifier drives NLC molecules and induces out-of-plane reorientation.

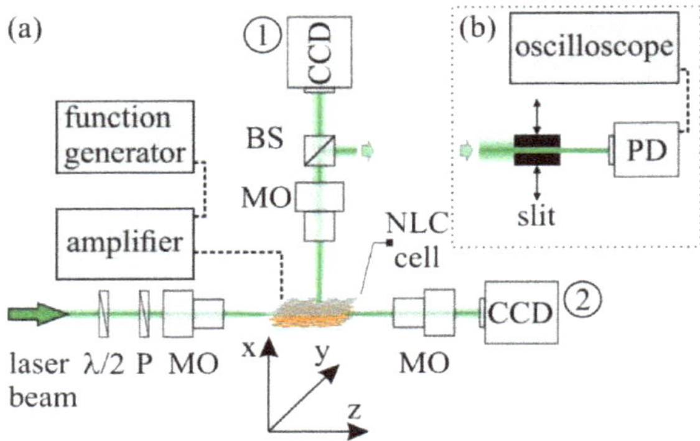

Figure 2. Sketch of the experimental setup showing: (**a**) part devoted to beam coupling and visualization of the beam propagation; (**b**) part for determining waveguide formation times. λ/2—halfwave plate, P—polarizer, MO—microscope objective, BS—beam splitter, CCD—digital camera, PD—photodiode.

To measure the formation times of waveguides induced by straight electrodes, a part of light collected by the first CCD camera is redirected to the photodetector (PD), schematically presented in Figure 2b. The amplitude of the electric signal recorded by the oscilloscope is proportional to incident light intensity. Without driving voltage applied to the electrodes of the NLC cell, the beam diffracts; therefore, a weak electric signal (mostly noise-related) is recorded by oscilloscope. When the beam is propagated within a waveguide, some light is scattered out of the propagation plane. Then, the light is collimated by the microscope objective reaching the photodiode. Therefore, an electric signal corresponding to a high state is recorded. To increase the dynamic range between a high and a low state, a mechanical slit that clips part of the beam is used.

On bare glass and metallic surfaces, NLC molecules are oriented randomly (in-plane, *xy*-plane); however, they remain parallel to the glass surface [28]. Under the influence of an external electric field (along the *x*-axis), NLC molecules (with positive dielectric anisotropy) start to reorient (out of plane, along the *x*-axis) and align along the electric field, i.e., towards homeotropic texture. This means that the initial molecular alignment (orientation in *xy*-plane) is not necessary to induce NLC waveguides (for TM-polarized beam) using an external electric field, as already presented by Fraunauhoffer [13]. In our work, the impact of alignment layers on electro-optically induced NLC waveguides was analyzed (e.g., driving voltages, formation times). In both configurations of NLC cells, the waveguide formation times and the driving voltages are comparable.

3. Results

3.1. Linear Waveguides

The light guiding within an electrically induced and controlled NLC waveguide is based on total internal reflection from two regions within the NLC layer with a different spatial distribution of the molecules [13,14,29]. When there is no voltage applied, the NLC molecules are uniformly oriented in a planar configuration (along the z-axis) over the whole volume of the sample. As the voltage increases (above the Freedericksz threshold value), molecules reorient out of plane towards the electric field lines to homeotropic orientation. Figure 3a presents the orientation of molecules in a uniform NLC cell characterized by the planar molecular alignment of a thickness equal to 12 μm, the same as the thickness of the analyzed NLC structures. The reorientation is induced by a uniform electric field; however, the final position of molecules depends, among others, on anchoring conditions, liquid crystal elasticity and applied voltage, as it is sketched for 6CHBT NLC for U = 1.2 V— black square, 1.5 V—gray circle, 2.0 V—light gray triangle, 2.5 V—magenta hexagon, 7.0 V—violet hexagon, and 20.0 V—purple star. As the molecules reorient out of plane, the refractive index for the input beam linearly polarized along with x increases according to $n_e(\theta) = n_o n_e/(n_e^2 \cos^2\theta + n_o^2 \sin^2\theta)^{-\frac{1}{2}}$, where: θ is the angle between the long axis and yz-plane, and n_0 and n_e are ordinary and extraordinary refractive indices, respectively, as is plotted in Figure 3b. In contrast, in the regions without electrodes, molecules maintain a planar orientation. Therefore, in a non-uniform electric field induced by structured electrodes, a transverse modulation of the refractive index is obtained. NLC cells with structured electrodes resemble a graded waveguide structure for a linearly polarized input beam with TM (along with x-axis) polarization.

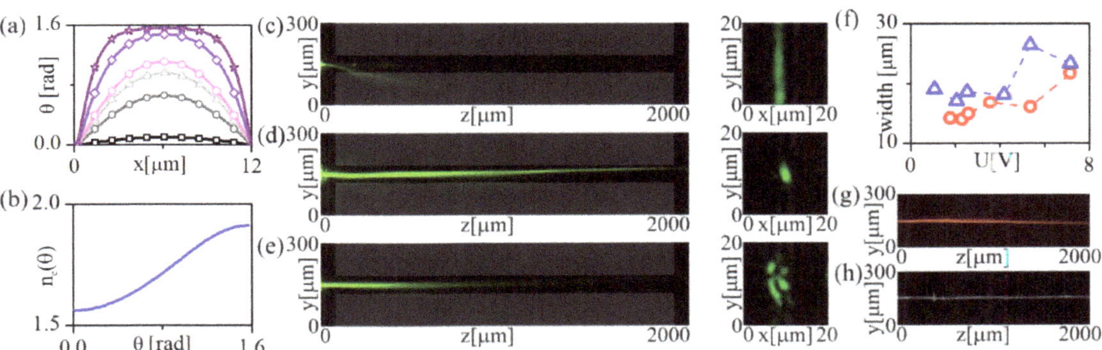

Figure 3. (**a**) Director distribution in NLC cell along the x-axis (cell thickness) as a function of the voltage (solid lines), plotted for U = 1.2 V—black (square), 1.5 V—gray (circle), 2.0 V—light gray (triangle), 2.5 V—magenta (hexagon), 7.0 V—violet (diamond), and 20.0 V—purple (star). (**b**) Refractive index for TM-polarized beam (λ = 532 nm) as a function of the orientation of molecules. (**c**–**e**) Experimental evidence of a (1D + 1) waveguide formation (electrode width 5 μm) and propagation of TM-polarized beam at λ = 532 nm. Beam evolution in yz-plane (left panel) and corresponding output profiles in xy-plane (right panel) (**c**) for U = 0 V, (**d**) for U = 2.3 V, and (**e**) for U = 7.0 V biasing voltage. (**f**) Mean beam width plotted for different voltages for 5 μm (red circles) and 10 μm (blue triangles) electrode widths. (**g**) Propagation of a TM-polarized beam at λ = 642 nm and (**h**) λ = 1064 nm, for U = 1.8 V and 5 μm electrode width.

Without an electric field, the NLC molecules maintain planar orientation $(\theta = 0^0)$ and the TM-beam propagates along the z-axis and diffracts as it does in isotropic media, with the refractive index n_0. For a driving voltage above the Freedericksz threshold value, the molecules reorient under the electrode, increasing the angle θ. As a result, the refractive index for the TM-polarized beam increases, and a waveguide channel is formed.

The laser beam propagation in the designed structure with a straight electrode is shown in Figure 3c–e. The linearly polarized input beam along the x-axis beam at λ = 532 nm is launched into the NLC sample in the region of the stripe-shaped electrode. Without the external voltage (U = 0), the light beam diffracts, increasing its width with propagation distance (Figure 3c). In the output plane (xy), this is visible as a broad illuminated region along the y-axis, confirming the beam diffraction in the yz-plane. Along the x-axis, the beam does not diffract significantly due to its limited thickness. Applying a voltage of (U = 2.3 V) causes a reorientation of the NLC molecules and thus an increase in the refractive index in this region, which consequently leads to the excitation of the waveguide channel (Figure 3d). The intensity distribution at the cell output is approximately Gaussian, characterized by partial ellipticity. The approximate diameter of the modal field is $w_x \cong 1.8$ μm and $w_y \cong 2.9$ μm, measured as FWHM values. Increasing the voltage up to 7 V widens the waveguide in both x and y directions due to further reorientation and saturation, and nonlocality of NLC molecules, respectively (Figure 3e). The beam output profile becomes irregular and corresponds to a mix of higher-order guiding modes. Therefore, the induced waveguide starts to support more and more higher-order modes.

The mean width of a beam as a function of biasing voltage on electrodes of a width equal to 5 and 10 μm is plotted and presented in Figure 3f. The beam is the narrowest for a biasing voltage of about 2 V, then, as the voltage increases, the mean width of a beam increases as well. Biasing voltage of 7 V results in irregular optical field distribution at the output. It indicates that propagation is realized as a superposition of higher-order modes.

An electrically induced waveguide also supports different wavelengths. The propagation of the visible beam at λ = 642 nm (Figure 3g) and infrared one λ = 1064 nm (Figure 3h) is recorded for a U = 1.8 V biasing voltage. Furthermore, NLC waveguides are not limited to straight geometry. An example is presented in Figure 4, where there are sketched arch-shaped electrodes of different sizes (left panel, Figure 4a) and a combination of three arch electrodes (left panel, Figure 4b). The propagation of a TM-polarized beam of a wavelength λ = 532 in a single arch-shaped channel and beam splitting at three different combined NLC arch-shaped electrically induced waveguides (driving voltage U = 2.3 V) are shown in Figure 4a,b (right panels). The possibility of beam splitting between more guiding channels makes it possible to design and perform novel photonics devices for integrated optical circuits.

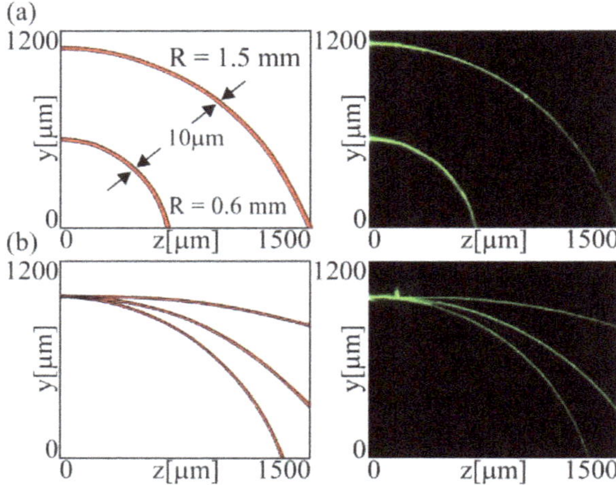

Figure 4. (**a**) Scheme of the curved geometry of the electrodes in NLC cell (left panel) and λ = 532 nm, TM-polarized beam propagation (right panel). (**b**) Combined three arch-shaped electrodes (left panel) and TM-polarized beam propagation at λ = 532 nm for U = 2.3 V.

3.1.1. Waveguide Formation Times

One of the important issues and also one of the significant drawbacks is the dynamics of the waveguide channel formation. Due to the fact that reorientation involves, in general, the rotation of the molecules, it is a relatively slow process. The relationship between the response time and the dynamic rotational viscosity γ and the cell thickness d is directly proportional: $t_r \sim \gamma d^2$. From this relation, the larger the dynamic rotational viscosity, the slower the response time, but the thickness is even more critical in determining the response time, as it is squared with the response time, which can be defined as the time that a certain region of an NLC cell takes to turn from on to off or vice versa. NLC with a low rotational viscosity favors a rapid switching using an electric field. For a specific NLC, the value of the viscosity cannot be changed freely; however, the electrical driving signal can be optimized to maximize the overall performance of the structure. In order to determine the formation time of the induced waveguide structure, the setup, as shown in Figure 1b, is used. The waveguide formation time is defined as the time from the moment the voltage is switched on until the transverse beam profile at a given propagation distance does not change and is more or less the same as the input beam profile. In an experiment, a more convenient way is to use a photodetector at the output of the waveguide. The criterion for the waveguide formation time that is measured is a time from beam coupling to the input to a moment when the electric signal at the photodetector reaches 0.9 times its maximum value. Generation of a waveguide channel requires a voltage just above the Frederiks threshold value. Such a voltage value forces the molecules to reorient, thus providing a change in refractive index $(n_e(\theta))$ sufficient to guide the light beam. However, in this case (relatively small voltage value), the reorientation time of molecules is on the order of hundreds of milliseconds (for the 12 μm thick cell).

The obtained NLC waveguide formation times are presented in Figure 5a. The reorientation time is estimated from a sequence of images taken by a CCD camera operating at a recording rate of 30 fps, i.e., the measurement resolution is 33 ms. No significant differences are observed in NLC cells with and without alignment layers at the bottom and top surfaces. However, the arrangement of all molecules towards the propagation direction seems to be a better approach to the reorientation process and slightly increases the response times of the molecules.

The reorientation rate directly depends on the amplitude of the driving voltage. For a driving voltage with an amplitude higher than the threshold value, the reorientation process occurs significantly faster than for voltages of the order of the threshold value. On the other hand, as shown already in Figure 3a, higher voltages are disadvantageous because they induce too high reorientation (and thus a change in refractive index) and promote/support higher-order modes. Therefore, we propose a solution of using two voltages: a high overdrive voltage to achieve fast molecule reorientations and a voltage to ensure single-mode waveguide maintenance. The duration time of this high amplitude driving voltage must be precisely determined. It is important to reduce the amplitude of a voltage to the optimum value needed for waveguide formation at a specific time. An overdriving voltage duration that is too short will not result in a sufficient reorientation angle to induce waveguide formation. On the other hand, when the duration of the overdriving voltage is too long, the molecules will reorient too much, and the induced waveguide becomes highly multimodal.

To determine the overdrive voltage duration, a typical configuration with two crossed polarizers and a planar NLC cell with the optical axis oriented at 45° to the polarizers' axis is used. Without an electric field, an NLC cell placed between crossed polarizers acts as a wave plate and brightens the field of view. As the voltage is switched on, the molecules start to reorient, and the phase delay between ordinary and extraordinary wave components changes. As a result, the intensity of a beam transmitted through the optical system varies in time. It stabilizes at a certain level depending on the final position of the molecules for a given voltage. Transmitted light intensity reaches the successive minima and maxima, finally reaching zero for the entirely reoriented molecules almost to a homeotropic texture. The number of oscillations increases with the NLC birefringence and cell thickness.

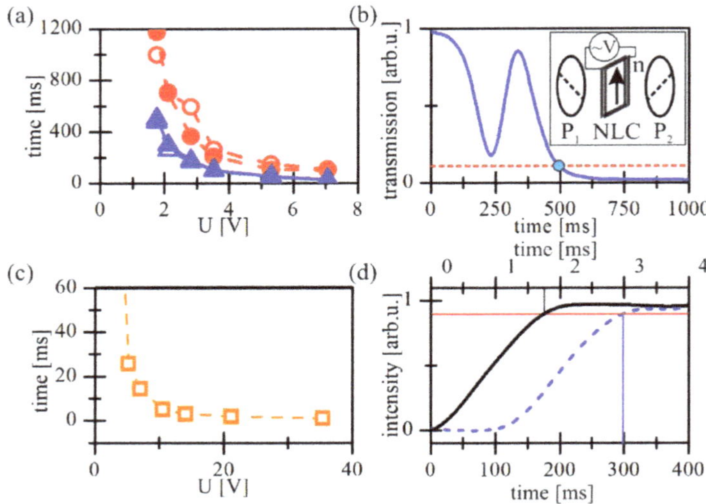

Figure 5. Experimental results on waveguide formation times and reorientation times. (**a**) Formation times as a function of voltage for 5 μm (red circles) and 10 μm (blue triangles) electrode widths. Open and filled markers refer to cells without and with an additional alignment layer, respectively. (**b**) Transmission through NLC cell vs. determined in the cross-polarizer configuration plotted for U = 2.1 V. The dotted red line denotes a decrease in the transmitted beam intensity to 10% of the maximum value. (**c**) Reorientation time as a function of a biasing voltage (orange squares). (**d**) Comparison of waveguide formation times for biasing voltage U = 2.1 V with (solid black line) and without 21.2 V overdriving voltage (dotted blue line). Intensity represents voltage rise time on the photodetector.

The oscillatory character of these changes for U = 2.1 V (the voltage value for which single-mode waveguide is formed) is presented in Figure 5b. In the measurements intended to determine the duration time of high overdriving voltage, a reorientation time is defined as the time from when the voltage is turned on until the intensity of the transmitted beam reaches 0.1 times its maximum value. From this, the reorientation times are determined for different overdriving voltages and plotted in Figure 5c. The line fitted to the experimental data is exponentially decreasing, and further increasing the voltage above a few tens of volts does not significantly reduce the reorientation time.

The waveguide induction time measurements are investigated as a function of overdriving voltage of duration, and amplitude presented in Figure 5b. The experiment is performed in the same experimental setup as in Figure 2a. A sinusoidal overdriving voltage of frequency 10 kHz and a precisely determined duration and amplitude, after which the voltage dropped to a value of U = 2.1 V, are applied to the NLC cell. The waveguide formation time is defined as the time that elapses from the voltage on until the electric signal generated on the photodetector reaches 0.9 times its maximum.

The application of a high overdrive voltage made it possible to shorten a switching time by two orders of magnitude. The best result is obtained for U_{ovr} = 21.2 V, and the formation time corresponds to 2.2 ms. The impact of a short duration voltage with high amplitude (2.2 ms; 21.2 V) is presented in Figure 5d.

3.1.2. Splitting Structures

Since the beam trajectory coincides with the shape of the electrodes, it is possible to design more complex electro-optical circuits designed for optical switching. The Y-splitter design is the simplest element to implement light splitting or switching. Figure 6a shows a top view image of the designed Y-splitter. It consists of three individual electrodes with a width of 10 μm each, namely, (1), (2) and (3), as marked on the photo, which allows for independent voltage control in each branch. The length of the first electrode to the forking (splitting) is equal to 600 μm; the angle between the second and third electrode is equal to 7° and the length of these electrodes is equal to 600 μm; thus, the distance between the output electrode (the "fork" width) is 50 μm. The voltages applied to the input electrode and two output electrodes are labelled as U_1, U_2, and U_3, respectively. The visible dark stripe perpendicular to the input electrode is used to supply the voltage and does not affect the beam propagation.

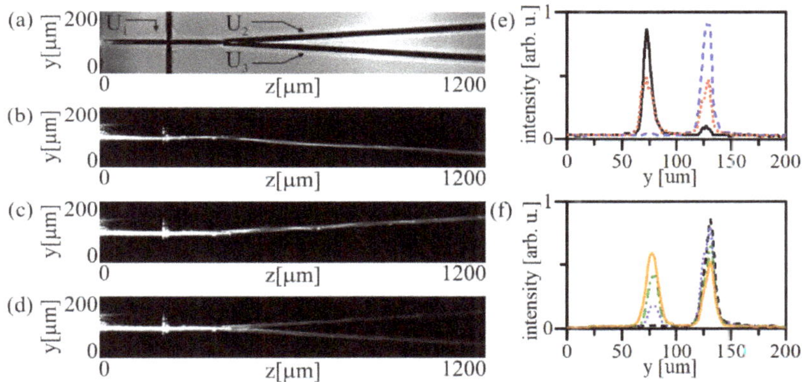

Figure 6. Electro-optic structure with three independent electrodes (with a width of 10 μm each) that forms an optical Y-junction for TM-polarized beam: (**a**) Photo of the designed structure illuminated by white light. (**b**–**d**) Beam propagation at λ = 1064 nm and (**e**) intensity cross-sections along the *y*-axis at z = 800 μm that corresponds to biasing voltage of 2.1 V applied to the electrodes: no. 1 and 3 (solid black line), no. 1 and 2 (dashed blue line) and no. 1, 2 and 3 (dotted red line), respectively. (**f**) Intensity cross-sections along the *y*-axis at z = 800 μm in case of biasing voltage of 2.1 V applied to electrodes 1 and 2, as a function of voltage at electrode no. 3: 0 V (dashed black line), 1.0 V (dotted blue line), 1.4 V (dash-dotted green line), 2.1 V (solid orange line).

Figure 6b–d shows the propagation result of linearly polarized TM infrared beam λ = 1064 nm in the fabricated structure. The first electrode, supplied with a voltage U_1 = 2.1 V is used to induce the input channel. Applying a voltage to the top electrode (U_2 = 2.1 V; U_3 = 0 V) induces the upper channel and the beam follows the direction (Figure 6b), while applying a voltage to the third electrode (U_2 = 0 V; U_3 = 2.1 V) switches the beam into the lower channel (Figure 6c). Applying the same voltage to all three electrodes ($U_1 = U_2 = U_3$ = 2.1 V) excites three waveguide channels in the medium, and the laser beam propagates along these channels (Figure 6c). The light is captured by a single input waveguide and then undergoes splitting/forking with approximately equal intensities due to the electrode geometry. Figure 6e,f shows the corresponding intensity distribution in the lower and upper channels at a distance of 800 μm for different voltages, illustrating the principle operation of the structure.

3.2. Nonlinear Waveguides

In a nonlinear case, an intense light beam in an NLC medium can propagate with a constant width at a propagation distance exceeding the Rayleigh range multiple times. For the TM-polarized beam, the generation of nematicon requires proper alignment direction. The strongest nonlinear response of the molecules is when molecules are oriented in the direction of the **k** vector. For a non-threshold reorientation of molecules in the presence of a light beam and avoiding the Freedericksz threshold effect, the initial pretilt of molecules is induced by an external electric field (Figure 7a). Without an electric field, molecules are subjected to a threshold, and, for the TM-polarized beam, diffraction is observed (Figure 7b). In the presence of an electric field, molecules tend to reorient towards the x-direction. The beam propagates in the form of spatial soliton (nematicon) and preserves its width at a distance of a few millimeters (Figure 7c).

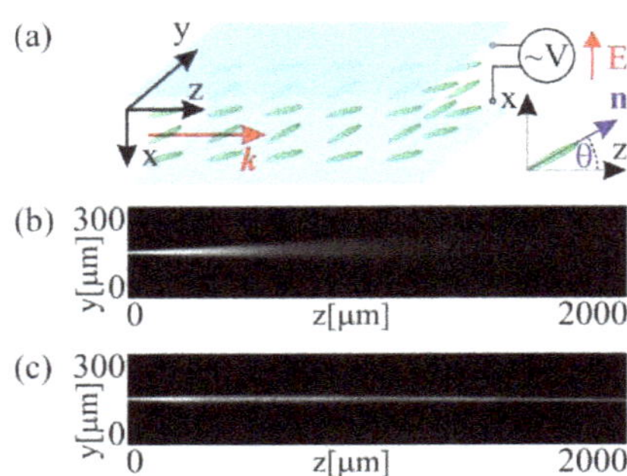

Figure 7. Experimental evidence of the light beam propagation in 6CHBT NLC cell, (λ = 1064 nm; optical power P = 1.5 mW): (**a**) The geometry of a cell and molecular orientation induced by an electric field of amplitude above the threshold value. (**b**) Diffraction observed in a *yz*-plane for beam propagation without biasing voltage. (**c**) Nematicons propagation for U = 1.6 V.

As described in the introduction part, in the nonlinear case, a light beam with sufficiently high intensity can propagate in NLC medium, with a constant width over distances many times the Rayleigh range, in the form of a so-called nematicon.

The alignment layers ensure planar orientation in the *yz*-plane with initial molecular anchoring at $\theta = 0$ with respect to the *z*-axis (Figure 7a). Indium/tin oxide (ITO) thin transparent electrodes deposited on the inner side of the NLC cell are applied to provide an adjustable low-frequency electric field for the out-of-*yz*-plane reorientation of the molecules. The NLC cell under consideration has been infiltrated with 6CHBT NLC and the linear polarized TM (along *x*-axis) light beam at λ = 1064 nm with optical power P = 1.5 mW has been focused to the waist of several micrometers and then launched into the NLC cell with its k vector oriented along the *z*-axis, as schematically presented in Figure 7a. Figure 7b presents an experimental photo of light beam propagation without an external electric field. As one can clearly see, the light beam diffracts. Without an external electric field, the reorientation is subjected to a threshold value much higher than the 1.5 mW. When a sufficiently high voltage is applied to the ITO electrodes, the NLC molecules are reoriented in the *xz*-plane, changing, thus, the director orientation. Indeed, by applying an electrical bias of 1.6 V (1 kHz), the self-trapped beam of TM-polarization is induced (Figure 7c). The beam propagates in the form of a nematicon and preserves its width at a distance of 2 mm.

Among other possibilities of controll light propagation in NLC structures [30–33], bycombining the propagation of the nematicon with the properties of structures with specially designed electrodes as described in Section 3.1, we obtain a configuration for changing the propagation direction of the nematicon in a Y-shaped geometry, as presented in Figure 8a. It is composed of one initial electrode of a width of 100 μm and two electrodes of a thickness equal to 50 μm with a short gap between them. The width of a cell is 50 μm. The amplitude of the driving voltage on each electrode can be controlled independently. The propagation direction of the nematicon between the upper (U_1) and lower (U_2) electrodes can be change by applying the voltage to the proper electrodes. To generate nematicon in the area designated by the initial electrode (U_0), a TM-polarized IR beam of initial width w_0 = 2.5 μm and optical power P = 1.5 mW is coupled to the middle part of the NLC cell. For the biasing voltage, the U_0 = 2.1 V beam forms a nematicon that propagates along the z-axis. Depending on the voltages U_1 and U_2, a nematicon is redirected to the upper or lower part of the NLC cell, as presented in Figure 8b,c. Trajectories of the nematicon that propagates in the case of U_0 = 2.1 V and U_1 = 2.1 V or U_2 = 2.1 V are plotted in Figure 8d. After propagation over a 1 mm distance, the spatial separation of a beam exceeds 300 mm, which may be advantageous for the application of the structure in integrated optical circuits. More NLC cells can be combined to obtain guiding/switching structures or decouple the beam from the NLC cell.

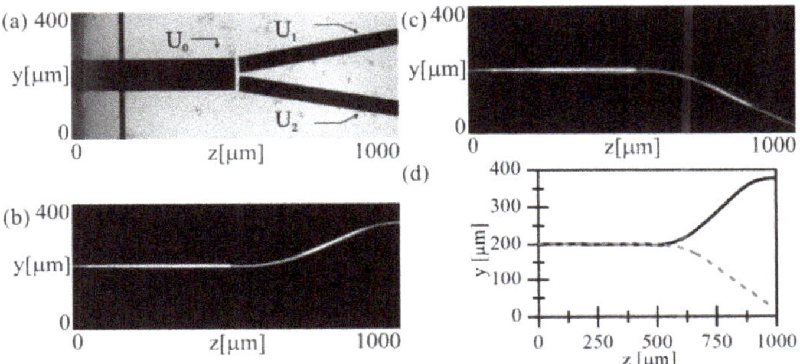

Figure 8. Combination of three independent powered electrodes of a width of 100 and 50 μm that formed an optical Y-junction for TM-polarized beam: (**a**) Electrodes within NLC cell. (**b**) Nonlinear propagation of a beam at λ = 1064 nm and optical power P = 1.5 mW for biasing voltage of $U_0 = U_1$ = 2.1 V and (**c**) $U_0 = U_2$ = 2.1 V. (**d**) Corresponding to (**b**,**c**) beam trajectories—black solid to light gray dashed line, respectively.

4. Perspectives

Expanding the structure by the additional electrode/electrodes can increase the number of output ports. The 1 × 3 Y-switching structure was prepared by the same method as the previously described 1 × 2 Y-shaped optical switches and is presented in Figure 9. It consists of four independent electrodes, one of which is devoted to inducing the pretilt of molecules in an initial channel, and the other three are responsible for changing the direction of nematicon propagation between upper (U_1), middle (U_2), and lower (U_3) output ports. Figure 9b shows a low-power signal beam (λ = 532 nm, TM-polarization) propagation, the width of which increases along with propagation distance. For a high-power TM-polarized beam (λ = 1064 nm, P = 1.6 mW) and driving voltage U_0 = 1.7 V, a nematicon is formed. A nematicon can be routed between three output ports depending on U_1, U_2, and U_3 driving voltages, as it is plotted in Figure 9c and shown in Figure 9d–f. A low-power co-polarized signal beam co-coupled with a high-power IR beam follows the nonlinearly induced waveguide channel presented in Figure 9g–i.

An interesting advantage of the proposed configuration is that the signal can be switched between the output channels not by changing the voltage value, but by using the phase shift of a sinusoidal electrical signal on the electrodes. A structure is analogous to the one described above and is shown schematically in Figure 10a. Based on the numerical calculations (presented in Figure 10b), it is seen that there is a considerable asymmetry in molecular reorientation out of the *yz*-plane in the proximity of the forked area when the out of phase electrical sinusoidal signal is applied to the upper electrode. In the qualitative numerical results, a theoretical value of reorientation angle in a cell midplane was found by solving the Euler–Lagrange equation, with the assumption of strong anchoring conditions at the boundaries (a planar alignment with an initial orientation of molecules at $\theta \cong 0$): $K_{11}\frac{d^2\theta}{dx^2} + \frac{1}{2}\varepsilon_0\Delta\varepsilon E^2 \sin 2\theta = 0$, where K_{11}—the splay elastic constant; E—amplitude of the electric field (above the threshold value) directed along the x-axis; $\Delta\varepsilon$—dielectric anisotropy; θ—reorientation angle of molecules out of the yz-plane. The voltages included in the calculation are $U_0 = U_2 = 2.0$ V. For the sake of simplicity, the counter phase of the voltage on the upper electrode is assumed to be $U_1 = -2.0$ V. In the considered situation, the reorientation of the molecules is asymmetric about the symmetry axis of the electrodes in the z-direction. Consequently, only one direction of nematicon propagation is preferred, resulting in switching the nematicons trajectory from the upper channel (when all-electrical signals are in phase) to the bottom channel (when the electric signal at the top electrode is out of phase with the others).

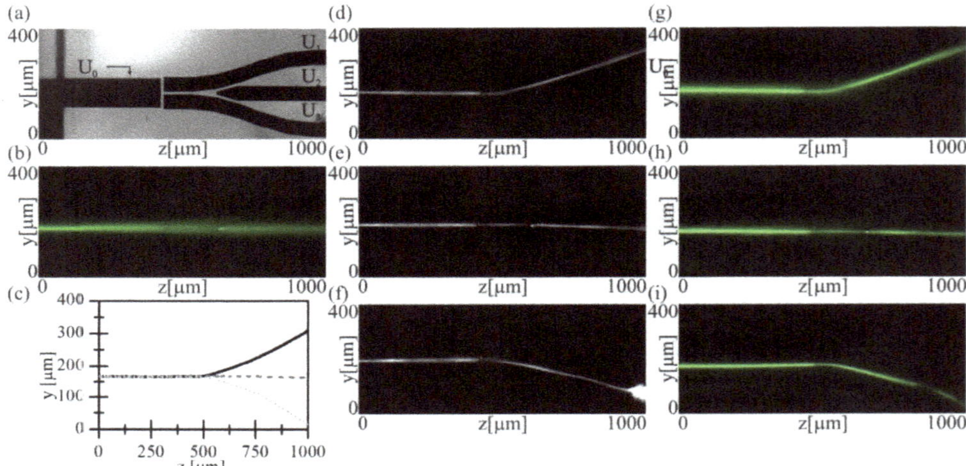

Figure 9. Combination of four independent controlled electrodes of a 100 μm width in the initial channel and 50 μm in the output channels formed an optical 1 × 3-junction for TM-polarized beam: (**a**) Electrodes within NLC cell. (**b**) Propagation of low-power signal beam ($\lambda = 532$ nm, TM-polarization). (**c**) Trajectories of self-induced nonlinear waveguides obtained for IR beam of optical power P = 1.6 mW, TM-polarization, $U_0 = 1.7$ V. (**d–f**) Nematicon propagation for driving voltage applied sequentially to the electrodes $U_1 = 1.7$ V, $U_2 = 1.7$ V and $U_3 = 1.7$ V. (**g–i**) Guided low-power signal beam ($\lambda = 532$ nm, TM- polarization) co-coupled collinearly with an infrared beam.

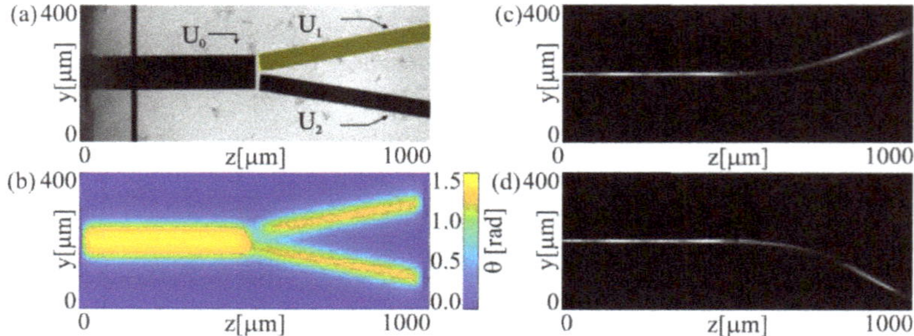

Figure 10. Redirection of nematicon in an optical Y-junction for TM-polarized beam using the phase shift of a driving voltage: (**a**) Electrodes within NLC cell, with a marked one with variable phase. (**b**) Asymmetry in the out-of-plane molecular reorientation when sinusoidal signal U_1 is out of phase with the U_0 and U_2 of the electrical signals—numerical calculations. (**c**) Experimental nonlinear beam propagation of a wavelength $\lambda = 1064$ nm and optical power P = 1.5 mW in case of $U_1 = 1.7$ V remains in phase and (**d**) out of phase to U_0 and U_2.

Experimental results presenting beam propagation and trajectory switching between two electrically induced channels due to a counter phase driving signal on the upper electrode (U_1) are presented in Figure 10c,d. The voltages of the sinusoidal signal (1 kHz) were $U_{0,1,2}$ = 1.7 V. A phase delay between the electric signal on the upper electrode (U_1) was varied, ranging from 0 to π. When all voltages remain equal, and in phase, the distribution of the effective refractive index for the TM-polarized IR beam in the forking section of the NLC cell is symmetrical, and the initial beam propagation direction is such that the beam propagates in the upper channel—Figure 10c. When the electric signal (U_1) was delayed by π, the effective refractive index in the area between electrodes in the initial part of a structure (U_0) and the upper one (U_1) decreases, and the beam propagation switches to a lower channel—Figure 10d. The optical power of a beam remains constant in both cases at P = 1.5 mW.

5. Conclusions

In this paper, electrically induced optical waveguides based on liquid crystals for a linearly polarized light beam have been presented. They operate by refractive index change induced by the reorientation of molecules under the influence of an external slowly varying electric field. Different geometries, both straight and curved waveguide structures as well as Y-shaped switches, have been presented. A method that significantly reduces the formation time of such waveguides (single milliseconds) has been proposed, which is undoubtedly an important step towards applications.

In the case of waveguides induced using both an external electric field and beam intensity (nonlinear case), independently controlled electrodes with a given geometry allow one to select the direction of signal (nematicon in this case) propagation by modifying the amplitude and phase. In particular, we aim to develop efficient methods to form liquid crystalline waveguide structures with the particular spatial distribution of the refractive index within in order to obtain functional elements typical for integrated optics (e.g., waveguides of different shapes, logic gates, Y-junction, switches, and so forth).

Author Contributions: U.A.L. conceived the original idea and wrote the paper. M.K. performed an experimental investigation. All authors have read and agreed to the published version of the manuscript.

Funding: This research was funded by National Centre for Research and Development, grant number LIDER/018/309/L-5/13/NCBR/2014.

Institutional Review Board Statement: Not applicable.

Informed Consent Statement: Not applicable.

Data Availability Statement: Not applicable.

Acknowledgments: We thank P. Jung (Warsaw University of Technology, Warsaw, Poland; University of Central Florida, Orlando, FL, USA) for his help in making the numerical calculations.

Conflicts of Interest: The authors declare no conflict of interest.

References

1. Tong, X.C. Electro-Optic Waveguides. In *Advanced Materials for Integrated Optical Waveguides*; Springer Series in Advanced Microelectronics; Springer International Publishing: Cham, Switzerland, 2014; Volume 46.
2. Courjal, N.; Bernal, M.; Caspar, A.; Ulliac, G.; Bassignot, F.; Gauthier-Manuel, L.; Suarez, M. Lithium Niobate Optical Waveguides and Microwaveguides. In *Emerging Waveguide Technology*; IntechOpen: London, UK, 2018. [CrossRef]
3. Findakly, T.; Suchoski, P.; Leonberger, F. High-quality LiTaO3 integrated-optical waveguides and devices fabricated by the annealed-proton-exchange technique. *Opt. Lett.* **1988**, *13*, 797–798. [CrossRef]
4. Eltes, F.; Caimi, D.; Fallegger, F.; Sousa, M.; O'Connor, E.; Rossell, M.D.; Offrein, B.; Fompeyrine, J.; Abel, S. Low-loss $BaTiO_3$–Si waveguides for nonlinear integrated photonics. *Acs Photonics* **2016**, *3*, 1698–1703. [CrossRef]
5. Qiu, F.; Spring, A.M.; Maeda, D.; Ozawa, M.A.; Odoi, K.; Otomo, A.; Aoki, I.; Yokoyama, S. A hybrid electro-optic polymer and TiO_2 double-slot waveguide modulator. *Sci. Rep.* **2015**, *5*, 8561. [CrossRef] [PubMed]
6. Kabanova, O.S.; Melnikova, E.A.; Olenskaya, I.I.; Tolstik, A.L. Electrically controlled waveguide liquid-crystal elements. *Tech. Phys. Lett.* **2014**, *40*, 598–600. [CrossRef]

7. Asquini, R.; Fratalocchi, A.; d'Alessandro, A.; Assanto, G. Electro-optic routing in a nematic liquid-crystal waveguide. *Appl. Opt.* **2005**, *44*, 4136–4143. [CrossRef] [PubMed]
8. Tabiryan, N.V.; Nersisyan, S.R. Large-angle beam steering using all-optical liquid crystal spatial light modulators. *Appl. Phys. Lett.* **2004**, *84*, 5145–5147. [CrossRef]
9. Yamaguchi, R.; Nose, T.; Sato, S. Liquid Crystal Polarizers with Axially Symmetrical Properties. *Jpn. J. Appl. Phys.* **1989**, *28*, 1730. [CrossRef]
10. Sirleto, L.; Coppola, G.; Breglio, G. Optical multimode interference router based on a liquid crystal waveguide. *J. Opt. A Pure Appl. Opt.* **2003**, *5*, S298. [CrossRef]
11. d'Alessandro, A.; Martini, L.; Civita, L.; Beccherelli, R.; Asquini, R. Liquid crystal waveguide technologies for a new generation of low-power photonic integrated circuits. In Proceedings of the Emerging Liquid Crystal Technologies X, San Francisco, CA, USA, 7–12 February 2015. [CrossRef]
12. Ntogari, G.; Tsipouridou, D.; Kriezis, E.E. A numerical study of optical switches and modulators based on ferroelectric liquid crystals. *J. Opt. A Pure Appl. Opt.* **2005**, *7*, 82. [CrossRef]
13. Costache, F.; Blasl, M. Optical switching with isotropic liquid crystals. *Opt. Photonik* **2011**, *6*, 29–31. [CrossRef]
14. Komar, A.A.; Tolstik, A.L.; Melnikova, E.A.; Muravsky, A.A. Optical switch based on the electrically controlled liquid crystal interface. *Appl. Opt.* **2015**, *54*, 5130–5135. [CrossRef] [PubMed]
15. Khoo, I.C. *Liquid Crystals: Physical Properties and Nonlinear Optical Phenomena*; Wiley: New York, NY, USA, 1995.
16. Peccianti, M.; De Rossi, A.; Assanto, G. Electrically assisted self-confinement and waveguiding in planar nematic liquid crystal cells. *Appl. Phys. Lett.* **2000**, *77*, 7–9. [CrossRef]
17. Beeckman, J.; Neyts, K.; Haelterman, M. Patterned electrode steering of nematicons. *J. Opt. A Pure Appl. Opt.* **2006**, *8*, 214–220. [CrossRef]
18. Assanto, G.; Peccianti, M.; Conti, C. Nematicons: Optical Spatial Solitons in Nematic Liquid Crystals. *Opt. Photonics News* **2003**, *14*, 44–48. [CrossRef]
19. Piccardi, A.; Alberucci, A.; Assanto, G. Nematicons and Their Electro-Optic Control: Light Localization and Signal Readdressing via Reorientation in Liquid Crystals. *Int. J. Mol. Sci.* **2013**, *14*, 19932–19950. [CrossRef]
20. Assanto, G. Nematicons: Reorientational solitons from optics to photonics. *Liq. Cryst. Rev.* **2018**, *6*, 170–194. [CrossRef]
21. Peccianti, M.; Conti, C.; Assanto, G. All-optical switching and logic gating with spatial solitons in liquid crystals. *Appl. Phys. Lett.* **2002**, *81*, 3335–3337. [CrossRef]
22. Peccianti, M.; Assanto, G. Signal readdressing by steering of spatial solitons in bulk nematic liquid crystals. *Opt. Lett.* **2001**, *26*, 1690–1692. [CrossRef]
23. Laudyn, U.A.; Piccardi, A.; Kwasny, M.; Karpierz, M.A.; Assanto, G. Thermo-optic soliton routing in nematic liquid crystals. *Opt. Lett.* **2018**, *43*, 2296–2299. [CrossRef]
24. Laudyn, U.A.; Piccardi, A.; Kwasny, M.; Klus, B.; Karpierz, M.A.; Assanto, G. Interplay of Thermo-Optic and Reorientational Responses in Nematicon Generation. *Materials* **2018**, *11*, 1837. [CrossRef]
25. Laudyn, U.A.; Kwaśny, M.; Karpierz, M.A.; Smyth, N.F.; Assanto, G. Accelerated optical solitons in reorientational media with transverse invariance and longitudinally modulated birefringence. *Phys. Rev. A* **2018**, *98*, 023810. [CrossRef]
26. Laudyn, U.A.; Kwaśny, M.; Sala, F.A.; Karpierz, M.A.; Smyth, N.F.; Assanto, G. Curved optical solitons subject to transverse acceleration in reorientational soft matter. *Sci. Rep.* **2017**, *7*, 12385. [CrossRef]
27. Dąbrowski, R. New Liquid Crystalline Materials for Photonic Applications. *Mol. Cryst. Liq. Cryst.* **2004**, *421*, 1–21. [CrossRef]
28. Kim, Y.; Lee, M.; Wang, H.S.; Song, K. The effect of surface polarity of glass on liquid crystal alignment. *Liq. Cryst.* **2018**, *45*, 757–764. [CrossRef]
29. Komar, A.A.; Kurochkina, M.A.; Melnikova, A.A.; Stankevich, A.I.; Tolstik, A.L. Polarization separation of light beams at the interface of two mesophases. *Tech. Phys. Lett.* **2011**, *37*, 704. [CrossRef]
30. Barboza, R.; Alberucci, A.; Assanto, G. Large electro-optic beam steering with nematicons. *Opt. Lett.* **2011**, *36*, 2725–2727. [CrossRef]
31. Kazak, A.A.; Melnikova, E.A.; Tolstik, A.L.; Mahilny, U.V.; Stankevich, A.I. Controlled diffraction liquid-crystal structures with a photoalignment polymer. *Tech. Phys. Lett.* **2008**, *34*, 861–863. [CrossRef]
32. Kazak, A.A.; Tolstik, A.L.; Mel'nikova, E.A. Controlling light fields by means of liquid-crystal diffraction elements. *J. Opt. Technol.* **2010**, *77*, 461–462. [CrossRef]
33. Sarkissian, H.; Serak, S.V.; Tabiryan, N.V.; Glebov, L.B.; Rotar, V.; Zeldovich, B.Y. Polarization-controlled switching between diffraction orders in transverse-periodically aligned nematic liquid crystals. *Opt. Lett.* **2006**, *31*, 2248–2250. [CrossRef]

Experimental results presenting beam propagation and trajectory switching between two electrically induced channels due to a counter phase driving signal on the upper electrode (U_1) are presented in Figure 10c,d. The voltages of the sinusoidal signal (1 kHz) were $U_{0,1,2}$ = 1.7 V. A phase delay between the electric signal on the upper electrode (U_1) was varied, ranging from 0 to π. When all voltages remain equal, and in phase, the distribution of the effective refractive index for the TM-polarized IR beam in the forking section of the NLC cell is symmetrical, and the initial beam propagation direction is such that the beam propagates in the upper channel—Figure 10c. When the electric signal (U_1) was delayed by π, the effective refractive index in the area between electrodes in the initial part of a structure (U_0) and the upper one (U_1) decreases, and the beam propagation switches to a lower channel—Figure 10d. The optical power of a beam remains constant in both cases at P = 1.5 mW.

5. Conclusions

In this paper, electrically induced optical waveguides based on liquid crystals for a linearly polarized light beam have been presented. They operate by refractive index change induced by the reorientation of molecules under the influence of an external slowly varying electric field. Different geometries, both straight and curved waveguide structures as well as Y-shaped switches, have been presented. A method that significantly reduces the formation time of such waveguides (single milliseconds) has been proposed, which is undoubtedly an important step towards applications.

In the case of waveguides induced using both an external electric field and beam intensity (nonlinear case), independently controlled electrodes with a given geometry allow one to select the direction of signal (nematicon in this case) propagation by modifying the amplitude and phase. In particular, we aim to develop efficient methods to form liquid crystalline waveguide structures with the particular spatial distribution of the refractive index within in order to obtain functional elements typical for integrated optics (e.g., waveguides of different shapes, logic gates, Y-junction, switches, and so forth).

Author Contributions: U.A.L. conceived the original idea and wrote the paper. M.K. performed an experimental investigation. All authors have read and agreed to the published version of the manuscript.

Funding: This research was funded by National Centre for Research and Development grant number LIDER/018/309/L-5/13/NCBR/2014.

Institutional Review Board Statement: Not applicable.

Informed Consent Statement: Not applicable.

Data Availability Statement: Not applicable.

Acknowledgments: We thank P. Jung (Warsaw University of Technology, Warsaw, Poland; University of Central Florida, Orlando, FL, USA) for his help in making the numerical calculations.

Conflicts of Interest: The authors declare no conflict of interest.

References

1. Tong, X.C. Electro-Optic Waveguides. In *Advanced Materials for Integrated Optical Waveguides*; Springer Series in Advanced Microelectronics; Springer International Publishing: Cham, Switzerland, 2014; Volume 46.
2. Courjal, N.; Bernal, M.; Caspar, A.; Ulliac, G.; Bassignot, F.; Gauthier-Manuel, L.; Suarez, M. Lithium Niobate Optical Waveguides and Microwaveguides. In *Emerging Waveguide Technology*; IntechOpen: London, UK, 2018. [CrossRef]
3. Findakly, T.; Suchoski, P.; Leonberger, F. High-quality LiTaO3 integrated-optical waveguides and devices fabricated by the annealed-proton-exchange technique. *Opt. Lett.* **1988**, *13*, 797–798. [CrossRef]
4. Eltes, F.; Caimi, D.; Fallegger, F.; Sousa, M.; O'Connor, E.; Rossell, M.D.; Offrein, B.; Fompeyrine, J.; Abel, S. Low-loss $BaTiO_3$–Si waveguides for nonlinear integrated photonics. *Acs Photonics* **2016**, *3*, 1698–1703. [CrossRef]
5. Qiu, F.; Spring, A.M.; Maeda, D.; Ozawa, M.A.; Odoi, K.; Otomo, A.; Aoki, I.; Yokoyama, S. A hybrid electro-optic polymer and TiO_2 double-slot waveguide modulator. *Sci. Rep.* **2015**, *5*, 8561. [CrossRef] [PubMed]
6. Kabanova, O.S.; Melnikova, E.A.; Olenskaya, I.I.; Tolstik, A.L. Electrically controlled waveguide liquid-crystal elements. *Tech. Phys. Lett.* **2014**, *40*, 598–600. [CrossRef]

7. Asquini, R.; Fratalocchi, A.; d'Alessandro, A.; Assanto, G. Electro-optic routing in a nematic liquid-crystal waveguide. *Appl. Opt.* **2005**, *44*, 4136–4143. [CrossRef] [PubMed]
8. Tabiryan, N V.; Nersisyan, S.R. Large-angle beam steering using all-optical liquid crystal spatial light modulators. *Appl. Phys. Lett.* **2004**, *84*, 5145–5147. [CrossRef]
9. Yamaguchi, R.; Nose, T.; Sato, S. Liquid Crystal Polarizers with Axially Symmetrical Properties. *Jpn. J. Appl. Phys.* **1989**, *28*, 1730. [CrossRef]
10. Sirleto, L.; Coppola, G.; Breglio, G. Optical multimode interference router based on a liquid crystal waveguide. *J. Opt. A Pure Appl. Opt.* **2003**, *5*, S298. [CrossRef]
11. d'Alessandro, A.; Martini, L.; Civita, L.; Beccherelli, R.; Asquini, R. Liquid crystal waveguide technologies for a new generation of low-power photonic integrated circuits. In Proceedings of the Emerging Liquid Crystal Technologies X, San Francisco, CA, USA, 7–12 February 2015. [CrossRef]
12. Ntogari, G.; Tsipouridou, D.; Kriezis, E.E. A numerical study of optical switches and modulators based on ferroelectric liquid crystals. *J. Opt. A Pure Appl. Opt.* **2005**, *7*, 82. [CrossRef]
13. Costache, F.; Blasl, M. Optical switching with isotropic liquid crystals. *Opt. Photonik* **2011**, *6*, 29–31. [CrossRef]
14. Komar, A.A.: Tolstik, A.L.; Melnikova, E.A.; Muravsky, A.A. Optical switch based on the electrically controlled liquid crystal interface. *Appl. Opt.* **2015**, *54*, 5130–5135. [CrossRef] [PubMed]
15. Khoo, I.C. *Liquid Crystals: Physical Properties and Nonlinear Optical Phenomena*; Wiley: New York, NY, USA, 1995.
16. Peccianti, M., De Rossi, A.; Assanto, G. Electrically assisted self-confinement and waveguiding in planar nematic liquid crystal cells. *Appl. Phys. Lett.* **2000**, *77*, 7–9. [CrossRef]
17. Beeckman, J.; Neyts, K.; Haelterman, M. Patterned electrode steering of nematicons. *J. Opt. A Pure Appl. Opt.* **2006**, *8*, 214–220. [CrossRef]
18. Assanto, G.; Peccianti, M.; Conti, C. Nematicons: Optical Spatial Solitons in Nematic Liquid Crystals. *Opt. Photonics News* **2003**, *14*, 44–48. [CrossRef]
19. Piccardi, A.; Alberucci, A.; Assanto, G. Nematicons and Their Electro-Optic Control: Light Localization and Signal Readdressing via Reorientation in Liquid Crystals. *Int. J. Mol. Sci.* **2013**, *14*, 19932–19950. [CrossRef]
20. Assanto, G. Nematicons: Reorientational solitons from optics to photonics. *Liq. Cryst. Rev.* **2018**, *6*, 170–194. [CrossRef]
21. Peccianti, M.; Conti, C.; Assanto, G. All-optical switching and logic gating with spatial solitons in liquid crystals. *Appl. Phys. Lett.* **2002**, *81*, 3335–3337. [CrossRef]
22. Peccianti, M.; Assanto, G. Signal readdressing by steering of spatial solitons in bulk nematic liquid crystals. *Opt. Lett.* **2001**, *26*, 1690–1692. [CrossRef]
23. Laudyn, U.A.; Piccardi, A.; Kwasny, M.; Karpierz, M.A.; Assanto, G. Thermo-optic soliton routing in nematic liquid crystals. *Opt. Lett.* **2018**, *43*, 2296–2299. [CrossRef]
24. Laudyn, U.A.; Piccardi, A.; Kwasny, M.; Klus, B.; Karpierz, M.A.; Assanto, G. Interplay of Thermo-Optic and Reorientational Responses in Nematicon Generation. *Materials* **2018**, *11*, 1837. [CrossRef]
25. Laudyn, U.A.; Kwaśny, M.; Karpierz, M.A.; Smyth, N.F.; Assanto, G. Accelerated optical solitons in reorientational media with transverse invariance and longitudinally modulated birefringence. *Phys. Rev. A* **2018**, *98*, 023810. [CrossRef]
26. Laudyn, U.A.; Kwaśny, M.; Sala, F.A.; Karpierz, M.A.; Smyth, N.F.; Assanto, G. Curved optical solitons subject to transverse acceleration in reorientational soft matter. *Sci. Rep.* **2017**, *7*, 12385. [CrossRef]
27. Dąbrowski, R. New Liquid Crystalline Materials for Photonic Applications. *Mol. Cryst. Liq. Cryst.* **2004**, *421*, 1–21. [CrossRef]
28. Kim, Y.; Lee, M.; Wang, H.S.; Song, K. The effect of surface polarity of glass on liquid crystal alignment. *Liq. Cryst.* **2018**, *45*, 757–764. [CrossRef]
29. Komar, A.A.; Kurochkina, M.A.; Melnikova, A.A.; Stankevich, A.I.; Tolstik, A.L. Polarization separation of light beams at the interface of two mesophases. *Tech. Phys. Lett.* **2011**, *37*, 704. [CrossRef]
30. Barboza, R.; Alberucci, A.; Assanto, G. Large electro-optic beam steering with nematicons. *Opt. Lett.* **2011**, *36*, 2725–2727. [CrossRef]
31. Kazak, A.A.; Melnikova, E.A.; Tolstik, A.L.; Mahilny, U.V.; Stankevich, A.I. Controlled diffraction liquid-crystal structures with a photoalignment polymer. *Tech. Phys. Lett.* **2008**, *34*, 861–863. [CrossRef]
32. Kazak, A.A.; Tolstik, A.L.; Mel'nikova, E.A. Controlling light fields by means of liquid-crystal diffraction elements. *J. Opt. Technol.* **2010**, *77*, 461–462. [CrossRef]
33. Sarkissian, H.; Serak, S.V.; Tabiryan, N.V.; Glebov, L.B.; Rotar, V.; Zeldovich, B.Y. Polarization-controlled switching between diffraction orders in transverse-periodically aligned nematic liquid crystals. *Opt. Lett.* **2006**, *31*, 2248–2250. [CrossRef]

MDPI
St. Alban-Anlage 66
4052 Basel
Switzerland
Tel. +41 61 683 77 34
Fax +41 61 302 89 18
www.mdpi.com

Crystals Editorial Office
E-mail: crystals@mdpi.com
www.mdpi.com/journal/crystals

www.ingramcontent.com/pod-product-compliance
Lightning Source LLC
LaVergne TN
LVHW070622170726
843515LV00004B/158
* 9 7 8 3 0 3 6 5 6 3 7 8 7 *